Spies in the Sky

Surveillance Satellites in War and Peace

Pat Norris

Spies in the Sky

Surveillance Satellites in War and Peace

 Springer

Published in association with
Praxis Publishing
Chichester, UK

Mr Pat Norris
Byfleet
Surrey
UK

SPRINGER–PRAXIS BOOKS IN SPACE EXPLORATION
SUBJECT *ADVISORY EDITOR*: John Mason, M.Sc., B.Sc., Ph.D.

ISBN 978-0-387-71672-5 Springer Berlin Heidelberg New York

Springer is part of Springer-Science + Business Media (springer.com)

Library of Congress Control Number: 2007929435

Cover design: Jim Wilkie
Project management: Originator Publishing Services Ltd, Gt Yarmouth, Norfolk, UK

Printed on acid-free paper

Contents

Preface

This is a story of intelligence estimates of weapons of mass destruction being wildly inflated, and of politicians using these inflated estimates to win elections. It's about the development of systems to detect weapons of mass destruction during which scientists have their careers destroyed by government cover-up. These could well be stories from the first 5 years of the 21st century, but in fact they come from the late 1950s. They are part of the story of the greatest contribution satellites have made to the world since the dawn of the space age.

The space age has now been with us for half a century, since Sputnik blasted into orbit on October 4th 1957 opening up an era of enormous excitement and anticipation—excitement at the discoveries and images returned by each new satellite; anticipation as mankind took the first tentative steps into the solar system.

In the short time since, satellites[1] have delivered many amazing achievements. The first ones were expensive and limited in performance. Now they are the cost-effective way to collect weather information or deliver TV to homes or tell you your position. Satellites are truly part of our everyday lives.

The greatest achievement of those first 50 years was clearly Neil Armstrong and his fellow Americans landing on the moon in 1969–1972—many people would say. Or was it the pictures from Jupiter, Saturn, and the surfaces of Mars, Venus, and Titan, changing forever our image of Venus from beauty to beast, and of Mars from bloody warrior to future tourism destination? Or surely the detection of global change including deforestation in the Amazon and shrinkage of the Antarctic ice cap must rank as the most important achievement of satellites?

Personally, I rank highly the role of telecommunication satellites in shrinking the globe. Thanks to these satellites the plight of famine victims in Africa, of earthquake and flood victims in Asia, and of war casualties in Iraq and Afghanistan is brought into our living rooms as it happens. The statements, actions, and follies of politicians

* Sometimes a "satellite" should be called a spacecraft, as explained in Chapter 2.

and celebrities on another continent can be watched as if next door. Sporting events that unite the world are witnessed by billions as they happen. Thanks to satellites, we now take for granted the ability to phone the other side of the world without having to pay half a week's salary for a crackly line. And while developed countries increasingly rely on undersea cables for inter-continental communications, Third World and isolated countries—in the Pacific Ocean, for example—are totally dependent on satellites for communication with the rest of the world. Not to mention seafarers and aviators for whom satellites are the only reliable form of communication when far from land.

So, we could certainly debate the most important contribution satellites have made to our life since 1957. In this book I hope to persuade you that the accolade as the most crucial role played by satellites since Sputnik was none of the achievements mentioned above, but was instead the relatively unsung role played by American and Soviet spy satellites in helping to prevent a nuclear holocaust during the Cold War.

The origins of this book lie in my own career. I have had the good fortune to work in two dynamic industries for the last 40 years—computers and satellites. Computers are at the heart of all major hi-tech endeavors these days, and through that fact I have had the opportunity to participate in several of the most exciting space programs of the first space half-century including the Hubble Space Telescope, the Apollo moon landings, Europe's first weather satellite, the landing of Huygens on Titan, the failed landing of Beagle 2 on Mars, the early telecommunications satellites, and the latest positioning satellites. Whether that experience makes me any more qualified to choose the most important space contribution of the half century I leave for you to decide.

Many people have helped me in the preparation of this book—one of the main challenges being to not impact my day job in the Space Division of LogicaCMG. My wife first encouraged me to write my opinions down. Others who have helped include (in alphabetical order) Richard Blott, Mike Cutter, Bob Kelley, Bill Levett, Martin Littlehales, John Mason, Ian Pryke, and Nick Veck. My publisher Clive Horwood has been a constant source of encouragement. Special thanks to David Harland and to Dwayne Day who provided many of the images. To these and others whose contribution has been important I offer my thanks, but I accept that the opinions and (hopefully rare) errors in the book are mine alone.

I have tried to attribute copyrights for the images used where they were evident. If anyone wishes to claim an image, I will happily amend the appropriate caption in the next edition of the book.

Pat Norris
July 2007

For Amy, Ciarán, and Valerie

Figures

Front cover

Rear cover

Abbreviations and acronyms

ABM	Anti-Ballistic Missile
AEHF	Advanced Extremely High Frequency
AIRSS	Alternative Infrared Satellite System
ASAT	Anti-SATellite [missile]
BBC	British Broadcasting Corporation
CCD	Charge-Coupled Device
CCTV	Closed Circuit TeleVision
CD	Computer Disk
CIA	Central Intelligence Agency
COMINT	COMmunications INTelligence
CSI	Crime Scene Investigation
DC	District of Columbia
Defcon	Defense Condition
DIA	Defense Intelligence Agency
DMS	Defense Meteorological Satellite
DoD	Department of Defense [US]
DSP	Defense Support Program
ELDO	European Launcher Development Organization
ELINT	ELectronic INTelligence
ESA	European Space Agency
EU	European Union
FIA	Future Imagery Architecture
GCHQ	Government Communications Head Quarters
GDP	Gross Domestic Product
GEOS	Geodetic Earth Orbiting Satellite
GPS	Global Positioning System
H-bomb	Hydrogen bomb
HDTV	High Definition TeleVision

Hex	Uranium hexafluoride
IAEA	International Atomic Energy Agency
ICBM	Inter-Continental Ballistic Missile
IGS	Information Gathering Satellite
INF	Intermediate-range Nuclear Forces [Treaty]
IRBM	Intermediate-Range Ballistic Missile
IRS	Indian Remote Sensing [satellites]
JPL	Jet Propulsion Laboratory
KH	KeyHole
LBJ	Lyndon B. Johnson
Li	Lithium
MAD	Mutually Assured Destruction
MFN	Most Favored Nation
MHz	MegaHertz
MIRV	Multiple Independently-targeted Reentry Vehicle
MIT	Massachusetts Institute of Technology
MoD	Ministry of Defence [UK]
MOL	Manned Orbiting Laboratory
MOX	Mixed-OXide
MRBM	Medium-Range Ballistic Missile
Musis	Multinational Space-based Imaging System
N-POESS	National Polar-orbiting Operational Environmental Satellite System
NASA	National Aeronautics and Space Administration
NATO	North Atlantic Treaty Organization
NCIS	Navy Criminal Investigation Service
NGO	Non-Governmental Organization
NOAA	National Oceanographic & Atmospheric Administration
NORAD	NORth American Aerospace Defense Command
NPT	Nuclear non-Proliferation Treaty
NRO	National Reconnaissance Office
NSA	National Security Agency [US]
OTH	Over The Horizon [radar]
Pan-STARRS	Panoramic Survey Telescope and Rapid Response System
PC	Personal Computer
Pu	Plutonium
RORSAT	Radar Ocean Reconnaissance SATellite
RPM	Revolutions Per Minute
SALT	Strategic Arms Limitation Treaty
SAM	Surface-to-Air Missile
SAOCOM	SAtellites for Observation and COMmunication
SAR	Synthetic Aperture Radar; Search & Rescue
SBIRS	Space-Based Infra-Red System
SDI	Strategic Defense Initiative
SDS	Space Data System

SIGINT	SIGnal's INTelligence
SLBM	Submarine-Launched Ballistic Missile
SLR	Single Lens Reflex
SPOT	Satellite Pour Observation de la Terre
SR	Space-based Radar
START	STrategic Arms Reduction Treaty
TAT	Trans-Atlantic Telephone
TNT	TriNitroToluene
TRW	Thompson Ramo Wooldridge
TSF	Télécoms Sans Frontières
TV	TeleVision
U	Uranium
UAV	Unmanned Aerial Vehicle
UK	United Kingdom of Great Britain & Northern Ireland
UN	United Nations
US/USA	United States of America
WW-II	World War II

1

Sputnik

INTRODUCTION

It was a bolt from the blue. Since World War II, the USA had dominated the non-communist world—socially and politically as well as economically. In 1950 the USA produced 40% of gross world product, the global equivalent of gross domestic product. The only country to have ever come close to that dominance had been Britain with roughly 20% of gross world product after the Napoleonic wars in about 1820.[1]

Then on October 4th 1957 the Soviets launched Sputnik (Figure 1) into orbit, demonstrating leadership in a hi-tech domain that most of the rest of the world had assumed resided in the USA.

The effect was electric, especially in the USA. The communists had been success-fully resisted in Berlin, Korea, and the Formosa Straits. It seemed that American military power could assert control anywhere in the world. Then overhead the heart-land of America itself came a shining communist star. This Sputnik transmitted radio signals, but perhaps the next one would carry a nuclear bomb! And America had no answer to it!

The surprise was unwelcome not least because of the echoes it triggered of Pearl Harbor 16 years before, when Japan sank much of the American Pacific fleet with a surprise aerial attack on the naval base of that name in Hawaii (Figure 2). Pearl Harbor brought about America's entry into World War II, so any similar event would evoke strong memories—only the Al Qaeda suicide plane crashes on 9/11 compare in the effect Pearl Harbor had on the American communal psyche.

And there had been other recent surprises—the explosion of the first Soviet atomic bomb in 1949 and the Chinese entry into the Korean War in 1950 being the two most embarrassing. As we will see later the American government reacted to

---

[1] Kennedy (1989) pp. 190, 475.

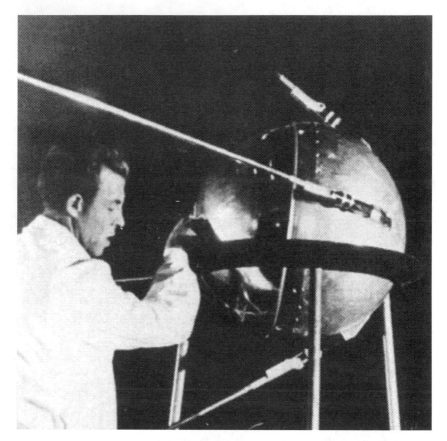

Figure 1. Sputnik: the world's first artificial satellite was placed in orbit on October 4th 1957.

these earlier surprises by investing heavily in spy planes such as the U-2, and had authorized development of a spy satellite two years before Sputnik. In fact, a successful Soviet long-range missile launch in August 1957 had alerted the American intelligence agencies to the ability of the Soviets to orbit a satellite, but they (and President Eisenhower) underestimated the impact the event would have on public opinion.

There was a torrent of heart-searching analysis from every corner, not least from Werner Von Braun, the German inventor of the V2 that had pummelled England in 1944–1945, now domiciled in the USA. He had been advocating a US space program in a series of TV shows and magazine articles throughout the early 1950s.

As the former commander of Allied forces in Europe during World War II, President Eisenhower, or "Ike" as he was affectionately known everywhere except Germany, was a big fan of satellites provided they did something useful. He had approved the development of a spy satellite two years earlier because he recognized that they provided a way to determine the progress in missile and bomber developments in the Soviet Union. At the same time he had rejected a number of projects

Figure 2. Ships of the American Pacific fleet ablaze after the Japanese surprise attack on Pearl Harbor, Honolulu, December 7th 1941.

from Von Braun and those of like mind, because he (Ike) was adamantly against expensive Federal programs that had no practical objective.

He urged his countrymen to see Sputnik for the publicity stunt it was, and he assured them that the US led in missile technology. The US could have launched a satellite a year earlier. In September 1956 Von Braun's Jupiter rocket reached a height of 1,100 km and a speed of 21,000 km/h. The fourth stage, instead of containing a rocket to give it the extra push to 29,000 km/h and into orbit, was filled with sand on orders from the top to prevent Von Braun "accidentally" placing it in orbit. Eisenhower's goal was to ensure that his spy satellite would be allowed to over-fly the Soviet Union. He was worried that if an overtly military rocket took the first US satellite into orbit the Soviets would cry foul and perhaps seek to have such satellites barred through the United Nations. Ike therefore wanted the first US satellite to have an American technical heritage rather than the German and military tinge that Von Braun's group automatically implied.

Eisenhower actually *welcomed* the launch of Sputnik as ensuring that the Soviets could not object when the US started flying its satellites over the Soviet Union before too long.[2] He had been pushing for a policy of "open skies" since 1955 in which both superpowers would allow the other's aircraft to make reconnaissance flights over

[2] DeGroot (2007) p. 57.

their territory to verify military preparedness. Of course, the Soviets, rightly, saw this as a cynical ploy whereby the US would allow the Soviets to observe the lead that the US had in bombers and missiles, and the US would watch to see if the Soviets were catching up.

Despite Ike's calming words, Sputnik unlocked a deep emotional response across America. The parallels with Pearl Harbor were explicitly voiced by many commentators. Edward Teller, for example—one of the creators of America's hydrogen bomb—declared on TV that the event was an even greater disaster than the Japanese surprise attack 16 years before. Some Congressional leaders called for a special session of Congress to debate the perceived massive Soviet lead in missile technology. Everyone understood that the carnage of the atomic bomb blasts that wiped out Hiroshima and Nagasaki to end World War II was now just a short missile trip away—should the Soviets so decide. Because of its great distance from its World War II enemies, Japan and Germany, America had been spared any significant attacks on its heartland during the War, but that accident of geography seemed now to have been consigned to history—and the public and politicians alike reacted emotionally.

Sputnik pushed Eisenhower's coldly logical arguments aside. The media engaged in a frenzy of alarm—the normally considered tones of the *New York Times* changed to ones of panic, advising its readers of a radical change in the military balance of power. The *Washington Post*, not noted for shrill headlines, wrote of a state of unpreparedness similar to 1941.

Politicians soon spotted an opportunity to hurt their opponents. Democratic leaders in both Houses of Congress took the opportunity to criticize the President and his party for allowing the Soviets to take the lead in space. "Our survival is at stake," said Senator Mansfield. Senate Majority Leader Lyndon B. Johnson (LBJ) warned that the Soviets would "soon be dropping bombs on us from space like kids dropping rocks onto cars from freeway overpasses"—demonstrating some gaps in his grasp of the physics involved. Upping the ante, he proclaimed that "control of space means control of the world … for the purposes of tyranny or for the service of freedom." LBJ was angling to be the Democratic Presidential candidate in 1960, and he kept the pressure up, calling Senate hearings on the space and missile gaps that started just six weeks after Sputnik. He demanded a national response similar to that of "the day after Pearl Harbor". Not to be outdone, House Speaker John McCormack worried that the US was headed for "national extinction" unless a response to Sputnik emerged. Another Presidential hopeful, Senator John F. Kennedy, jumped on the band wagon, proclaiming Sputnik "the most serious defeat the United States has suffered in many, many years." In the face of this cacophony of alarm, Ike's calmness looked like complacency.[3]

The Administration made a few attempts to attribute the Soviet success to the technology and people they had captured in Germany at the end of World War II— that is, "their Germans are better than our Germans." Not only did this line not go

---

[3] DeGroot (2007) pp. 50–72.

down too well with "our Germans"—Von Braun and his German colleagues in Huntsville—it did nothing to dampen public concern.

A political momentum emerged over-night for the USA to emulate the Soviet feat. Before long, the US Navy missile teams had as much funding as they could absorb to get a satellite in orbit. But Von Braun's Army team was still held back because of its militaristic connotations.

SPUTNIK-2 AND 3

In the Soviet Union and the rest of the communist block Sputnik was also a surprise, but one that was greatly welcomed. Like Eisenhower, Soviet Premier Nikita Khrushchev was unimpressed by Sputnik at first, seeing it as just a powerful rocket. The official communist newspaper, *Pravda*, carried the story of the launch as a small item on the front page without mentioning the name of the satellite.

That soon changed! Seeing the hysteria created around the world, *Pravda* two days later devoted almost the whole of the front page to Sputnik. The political impact of this space age coup was not lost on the leadership. World leaders who were normally antagonistic towards, or disinterested in, the Soviet Union congratulated Khrushchev on its success. Every opportunity was used to trumpet the technology leadership, and by implication the moral leadership, of the Soviets. The Soviets condescendingly assured the world that "there is no danger to any of the peoples of this world from this man-made moon".[4]

The surprises kept coming.

Sputnik 1 weighed about 83 kg or nearly 200 lb, and only one month later the Soviets launched Sputnik-2 weighing over half a ton. Sputnik-2 also contained a dog, Laika, which lived four days in space until a spacecraft failure caused her death. The ability of the Soviets to put a second satellite into orbit so soon after the first was impressive—and scary. The ability to carry a living creature suggested that "where dog goes, man will follow", and the thought of a Soviet person flying over the US unimpeded made Americans queasy.

Then just 6 months after that, in May 1958, the massive 1.5-ton Sputnik-3 was orbited, demonstrating—if that were still needed—that the Soviets had massive rocket capability. Because Soviet events took place in total secrecy they were able to suppress news of their failed launches and only report the successes. In contrast, America's launches took place in the full glare of publicity with the world's press and broadcasters present. The perception that the US failed a lot and that the Soviets were always successful fueled the public paranoia.

The Soviet space successes were spectacular but little else. But they were spectacular. And this only heralded the real eye opener, the launch of the first man into space in 1961. Cosmonaut Yuri Gagarin became a household name across the globe when he orbited the earth on April 12th that year (Figure 3). His charm and good looks enhanced his ability to play the role of communism's ambassador to the world.

[4] DeGroot (2007) pp. 66–68.

Figure 3. Space as politics: Yuri Gagarin, the first man to orbit the earth, and Soviet Premier Nikita Khruschev on the dais in Moscow's Red Square, April 1961. Picture courtesy: British Interplanetary Society archives.

He was feted wherever he went, hosted by Royalty and Presidents in Western countries as well as in communist and neutral ones.

THE US FIGHTS BACK

Meanwhile, the USA was playing catch-up in this new "space race". The US Navy team had been building the Vanguard rocket since being selected to launch the first US satellite two years earlier in August 1955—much to the disgust of Von Braun and his Army colleagues. Vanguard had no other purpose than to launch a scientific satellite and—unlike Von Braun's Jupiter rocket—was not a modified missile, thus achieving the appearance of non-military motivation that Eisenhower wanted. Sputnik-2 with its canine passenger had such a powerful impact that Eisenhower gave in to the calls from all sides for the US to respond.

The first attempt by the Navy Vanguard team to emulate Sputnik in December 1957 failed with an explosion seconds after launch. There was an orgy of soul searching about how far behind America had fallen. The Soviet delegation to the United Nations offered financial aid to the US as part of a program of technical assistance to backward countries.

It wasn't until February 1st 1958, that one of Von Braun's US Army Jupiter rockets carried Explorer 1 into orbit. Unlike the massive Soviet Sputniks, Explorer 1 weighed in at just 14 kg.

These early flights illustrated some of the features inherent to each of the societies that created them. The Soviet launches took place in secrecy and were only announced to the public once they were successfully in orbit. The American launches, failures as well as successes, appeared live on television for all to see. The American openness had many advantages when the launches worked, but they were embarrassing for all concerned when they failed—as they did all too frequently in those first years. It was difficult for the engineers to work efficiently in the full glare of publicity, not least because of the immediate and copious criticism from the politicians that provided the funding.

The Soviets too were subject to criticism, albeit in private. Although taking place out of the limelight, the penalty for failure in the Soviet Union was inherently more severe than in the West. Chief Designer Sergei Korolev had spent several years in a concentration camp during the Stalin era. The Khrushchev regime of the late 1950s was more tolerant than under Stalin, but memories of the treatment meted out to dissenters and undesirables were still high in people's minds. Thus, Korolev didn't hesitate when Khrushchev demanded that Sputnik-2 be built and launched within a month of Sputnik 1 so that Khrushchev could trumpet its success to the Communist Party Congress. Khrushchev had been lukewarm about Korolev's Sputnik project until he saw the massive propaganda coup it provided after Sputnik 1 was launched.

The difference in openness was also apparent in other ways. Journalists received detailed briefings on every aspect of the American attempts, but the Soviets were parsimonious in providing information. And it wasn't always accurate—the most famous example was when Gagarin followed instructions and actually lied at his press conference by stating that he had landed in his capsule. In fact, the design had called for him to bail out of the capsule and descend on his parachute, which he had duly done. The lie was fabricated to ensure that the Soviets could claim the first launch and return of a man into space—ejecting in the atmosphere arguably disqualified the Soviets from claiming a successful return.

THE CHIEF DESIGNER

Perhaps the most bizarre aspect of Soviet secrecy was the anonymity of Korolev (Figure 4). The few references to him spoke only of "the Chief Designer", without naming him. The Soviet leaders were paranoid about American abilities to entice scientists to defect to the West, and refusing to name the head of their space program was part of their reaction. Other reasons had to do with Korolev's shady past (in communist eyes) including a spell in the gulag (concentration camp system) in 1938–1944. By hiding Korolev from the limelight, the Soviets missed a great opportunity to dethrone Von Braun as the world's leading publicly recognised rocket scientist. Korolev's achievements arguably exceeded Von Braun's in that he created not only the rockets (and missiles) but also the satellites that they carried into space, while Von Braun was largely excluded from the satellite business.

Figure 4. Sergei Korolev, the Chief Designer, May 1961, with some of the first group of cosmonauts; front row (L to R): Yuri Gagarin (first man in space), Korolev, Yevgeni Karpov (head of the cosmonaut corps); back row (L to R): Pavel Popovich (made two trips to space, 1962 and 1974), Grigori Nelyubov (expelled from cosmonaut corps in 1963 after drunken brawl, died 1966), Gherman Titov (second man to orbit the earth in August 1961, died 2000), Valeri Bykovsky (three space trips in 1963, 1976, and 1978). Picture courtesy: British Interplanetary Society archives.

"He had unlimited energy and determination, and he was a brilliant organiser," Khrushchev said in his 1974 memoirs, adding that "his reports were always models of clarity." [5]

Korolev's interest in rocketry began in the 1920s. According to the now almost legendary version of his life story, in 1929 he visited the generally accepted "father of rocketry", Konstantin Tsiolkovsky, who enthused him with the mission of building rockets to reach earth orbit. Tsiolkovsky died in 1953, four years short of his centenary and the fulfilment by Korolev of his dream. Before being sent to the gulag Korolev had led the development of rockets for the Soviet military. In the bizarre Soviet system, during World War II and still in prison, he was ordered to continue his rocket work—he even received the Medal for Valiant Labor while still a prisoner.

[5] DeGroot (2007) p. 55.

Korolev's death in 1966 was a great loss to the Soviet space efforts. Thereafter, they failed to achieve many of their goals, most notably the race to the moon. It wasn't until the softening of Soviet secrecy in the 1980s that his name became known in the West.

In death as in life, Korolev was surrounded by secrecy. He went to hospital for an exploratory abdominal operation, but because of his importance to the Soviet state the Health Minister himself felt compelled to undertake the operation. The Minister had not prepared as thoroughly as he should have and botched the operation, causing Korolev's death.[6] This story is attested to by several Soviet commentators but is not yet the official version—maintaining Korolev's veil of secrecy even now.

SUBSTANCE VERSUS HYPE

The much greater weight of the Soviet satellites was to some extent compensated for by the greater sophistication of the American ones. The Soviets missed the first major scientific discovery of the space age by virtue of not carrying instruments on Sputniks 1 and 2, and by a faulty tape recorder on Sputnik-3. The tiny American Explorer 1 carried a radiation detecting instrument (Geiger counter) that enabled its designer James Van Allen to discover the radiation belts that encircle the earth, and which are now known as the Van Allen belts. As Explorer 1 passed the high point in its elliptical orbit, 1,500 km above the earth, the instrument fell silent as if there was absolutely no radiation. Van Allen correctly interpreted this to mean that the instruments had become saturated because the radiation was too intense, thereby discovering the radiation environment that now bears his name, and becoming the first science superstar of the space age.

The technical heritage of the two nations was also subtly different. President Eisenhower was at pains to distinguish between the military missile and scientific satellite communities. On the assumption that the USA would be the first to launch a satellite, he wanted to avoid the Soviets perceiving it as a threat to their sovereignty. So he assigned responsibility for missiles to a different group than the one tasked with launching a satellite. A year after the Sputnik shock Eisenhower created NASA as a civilian agency responsible for non-military space activities, thus formalizing the split. Eisenhower was opposed to civilian space programs, seeing them as nothing but prestige exercises. He worried that they would drain away resources and talent from the military space program which he considered strategically important. He acceded to the formation of NASA under pressure from both parties in Congress, large parts of the media, and the scientific community.[7] In response to the same pressures, he also created the post of scientific adviser, and appointed the President of MIT, James Killian, to it.

The difference between public perception and actual achievement was already apparent a year after Sputnik—it would become even more so later. In 1958, 23

[6] Siddiqi (2003) p. 514.
[7] Ambrose (1984) p. 458.

launches were attempted of which 17 were by the US. Of the 6 Soviet launch attempts, 5 failed to reach orbit—Sputnik-3 in May was the only success. Of the 17 US launch attempts, 7 reached orbit and 10 failed. So on the face of it, the US had a 7 : 1 lead in 1958. The much maligned US Vanguard even made it into orbit in March. NASA had its first success ten days after opening its doors on October 1st with Pioneer 1 which set an altitude record of over 100,000 km, and even the Soviet heavy lift record was broken on December 18th with the launch of a 4-ton experimental communications satellite on the Atlas—a converted ICBM.

But the popular perception was the reverse—the 10 US failures made the head-lines, drowning the good news in images of rockets exploding on the launch pad at Cape Canaveral. Meanwhile, the Soviet failures were kept secret until many years later, so the only news about their exploits was of success.

This disjoint between public perception and space realities continued. In April 1960 the US had 10 satellites in orbit, the Soviets 2. That same year the US launched the first weather satellite that immediately improved weather forecasting, and the first navigation satellite that improved maritime navigation. In contrast, the Soviets launched a satellite in August containing two dogs which of course garnered much more attention than the worthy US achievements—at least the dogs got back safely this time. In total, in the three years after Sputnik, the US put 28 satellites in space, the Soviets 8. In 1961 the pattern was the same—the US put 20 satellites in orbit, the Soviets 5. A 1960 opinion poll in Europe showed that only 20% thought the US was now superior militarily to the Soviets—down dramatically from five years earlier.[8]

CIVIL–MILITARY

While Ike kept military and civil space separate, the Soviet satellite and missile programs were closely aligned, with Korolev overseeing both. In fact, initially Korolev had to lobby hard to have satellite projects tagged on to the Soviet missile developments. The Sputnik 1 launch came just 6 weeks after a successful test of the latest Soviet ICBM—and that missile success was crucial in gaining Korolev the final authorization to proceed with Sputnik 1.

The military–civil link became even more explicit when Korolev sought author-ization for the Gagarin manned flight. Korolev got approval for a manned flight in November 1958 but the funding needed to make it happen was not approved. Then the first pictures from space (from US weather satellites, as it happened) underpinned an urgent program to build spy satellites, which Korolev managed with difficulty to amend to include the phrase *"and a satellite designed for manned flight."*[9] Using the high priority of the spy satellite program, Korolev was able to press ahead with a manned program.

In the US, Sputnik gave a major push to the program to build spy satellites, which we will return to in Chapter 4. The Administration, however, kept this

[8] DeGroot (2007) pp. 99, 120.
[9] Kislyakov (2006) pp. 229–231.

initiative under a veil of secrecy and publicized only its civilian space activities. The Republicans would have liked to point to the developments of surveillance satellites to rebut the accusations of complacency thrown at them by their opponents at home and abroad, but Ike was adamant that it remain under wraps for military reasons. The Republicans suffered in the 1958 mid-term elections as a result, and again in the 1960 elections—the Democrats made much of a supposed missile gap that the Soviets had opened up over America as illustrated by Sputnik and leaks of selective intelligence findings to the media by defence officials seeking increased budgets—the leaked intelligence being restricted to those parts of the overall picture that supported the leaker's motivation, and suppressing the other parts. As we will see in Chapter 5 the first spy satellites soon showed that the missile gap was in America's favor, not the Soviet's.

END OF AN ERA

The effect of Sputnik 1 on the American public was intense even though the 1950s are often characterized as having been a decade of complacency. There was no Great Depression as in the 1930s, no World War II as in the 1940s, no Vietnam War and civil rights unrest as in the 1960s, no oil embargo and inflation as in the 1970s. Certainly for the majority of Americans the 1950s was a decade of stability and advancement. President Eisenhower and the Republican Party benefited from this, with electoral success in 1952 and 1956. However, nostalgia for the 1950s is probably only in reality for a period of four years—from the summer of 1953 (after a bitter and divisive 1952 election and the ending of the Korean War) to October 1957 and Sputnik.

Sputnik gave the Democrats a focal point for criticizing the Republicans (not Eisenhower, whose personal popularity remained consistently high): for the glaringly obvious short-term failing in satellites and missiles, for the assumed long-term failings in science and education, and for the loss of prestige. The American public lapped it up and gave the Democrats victories in the 1958 and 1960 elections. So Sputnik marked the end of an era of relative consensus in American society and politics, paving the way for the social upheavals of the 1960s, most notably in the civil rights arena.

Eisenhower at first tried to play down the significance of Sputnik, dismissing it as merely evidence that "they can hurl an object a considerable distance" and declaring that the USA led in missile research.[10] Sputnik convinced most people that the Soviets now had the capability to bomb America with impunity, and the American public didn't like it.

---

[10] Ambrose (1984) p. 429.

2

After 50 years—satellites in our daily lives

INTRODUCTION

The launch of Sputnik on October 4th 1957 was the start of the space age. In the intervening 50 years there have been several "where were you when . . .?" moments, to which we can all relate provided we are old enough. Most have been associated with disasters such as the shooting of President Kennedy in 1963 or the suicide terrorism of 9/11. These moments remain in our memories long after other events before and after have faded. We remember not only the earth-shattering event itself but our own actions and surroundings at the time. They are carved into our memories.

One of the good news events that falls into this category was Neil Armstrong stepping onto the moon's surface on July 20th (or 21st, depending on where you were in the world) 1969. His words "one small step for a man, one giant leap for mankind" are instantly recognizable, and revive the memories of that day (Figure 5). Surely then this Apollo 11 mission must rank as the greatest achievement of the first 50 years of the space age!

One of the surprising things about Armstrong's moment of history was that it was available for all to see on television as it happened (except in China, which kept the news from its citizens for several years). This wasn't science as we know it, where hours of laboring in a laboratory results in a carefully documented breakthrough. It was exploration of a kind, similar in a way to Columbus reaching America, but the knowledge of the moon we gained came not from the observations of the hardy astronauts but from scientific experiments they left on the moon, and above all from the samples of the moon they brought back to earth—where they *could* be analysed for hours in a laboratory. And both of these could have been achieved robotically as was demonstrated by the Soviets at the time.

So, the historic importance of Apollo is tempered by the knowledge that it was largely unnecessary from a scientific point of view. It was done because it *could* be done, not because it needed to be. It was done to demonstrate to the world that

Figure 5. Apollo 11; Neil Armstrong takes a photo of Buzz Aldrin, Tranquility Base, July 20th 1969. Credit: NASA.

America had the technological know-how to achieve almost any goal—the importance of the television element was to deliver that message to the world. The Apollo astronauts, too, were a major part of the demonstration. They took on the mantle hitherto held by Yuri Gagarin, personalizing an otherwise abstract technical achievement in a way that people could relate to.

Apollo was designed to restore confidence to a shaky American public reeling from the shock of Sputnik. During the 1960s, as the American Mercury and Gemini manned space programs progressed, that confidence was indeed largely restored. The moon landing within the timetable set by President Kennedy in 1961 set the seal on that confidence-building exercise. But the world had moved on in the 12 years since Sputnik. America was embroiled in a war in Vietnam in which its undoubted superior

Figure 6. President John F. Kennedy proposes the Apollo moon program to Congress flanked by Vice President Lyndon Johnson (left) and House Speaker Sam Rayburn, May 25th 1961. Credit: NASA.

technical and economic prowess seemed powerless against the commitment and tenacity of the Viet Cong. Apollo 11, it turns out, was a bit of a hollow victory, demonstrating unrivaled competence in an abstract and scientific domain, but leaving more mundane earth-bound problems unresolved—not only in Vietnam, but at home in such things as segregation, urban ghettos, and poverty.

The decision to send Americans to the moon was also not as visionary and heroic as one might think. Kennedy's speechwriters certainly made it sound so, and captured the mood of the nation with his ringing phrase to Congress on May 25th 1961 (Figure 6): "I believe that this nation should commit itself to achieving the goal, before this decade is out, of landing a man on the moon and returning him safely to the earth." However, the steps leading up to this decision were tactical and short-term, and not at all visionary.

Sputnik had thrown Eisenhower's carefully crafted plans into disarray, and, as historian Gerard DeGroot points out, Gagarin was Kennedy's Sputnik. Let's look at the sequence of events in those few weeks of 1961. It started on April 12th with Yuri Gagarin, a Soviet citizen, becoming the first human to orbit the earth. All the alarms of Soviet missile leadership triggered by Sputnik in 1957 were sounded again—

America in decline, Soviets have a lead in missile technology, etc. Five days later, April 17th, Cuban exiles backed by the CIA invaded Cuba at Bahia de los Conchos (Bay of Pigs) and were crushed by Fidel Castro's forces. Kennedy's credibility, popularity, and authority took a dive. On April 25th an Atlas rocket testing the Mercury capsule in which US astronauts would soon ride blew up 40 seconds after launch. Congressional leaders felt frustrated and angry and demanded that NASA "beat the Russians". Kennedy needed something to take the spotlight off this string of political failures.

The moon-landing idea had been studied by NASA for several months and now emerged as the way to respond to these calls for action. Kennedy himself was against it on grounds of cost—the total Federal budget that year was $94 billion, and NASA chief James Webb put the cost of a moon trip at $20–40 billion. The idea got support from a surprising quarter—the Department of Defense. Defense Secretary Robert McNamara gave it strong support on the grounds that it would give the aerospace industry badly needed work without coming out of the defense budget.

The final event that persuaded Kennedy to commit to the Apollo program (as the moon-landing project became known) came on May 5th. The first Mercury capsule containing a human made it into space—not into orbit, mind you, but 187 km up then straight back down. The astronaut who was inside the capsule, Alan Shepard, became an immediate hero to the nation. Forty-five million watched it on TV. A quarter of a million turned out for his ticker tape parade in New York. The public reaction was not lost on Kennedy—nor on Congress as he discovered in soundings taken after the Shepard flight. The human element in a space project trumped all the science and technical wizardry when it came to public (and therefore political) interest. Thus, it was in the full knowledge that Congress was enthusiastically, even impatiently, waiting for his moon proposal that Kennedy announced the plan to land men (Americans) on the moon "before the decade is out".

From the moment Alan Shepard squeezed into the Mercury capsule in 1961 ("spam in a can" as test pilot legend Chuck Yeager put it[11]), astronauts have been the focus of public interest in space. One of Shepard's six Mercury program colleagues, Wally Schirra, bought his Maserati from Brigitte Bardot, epitomizing the film star aura the job brought with it. They became the comic book heroes that the public wanted.[12] In the light of Kennedy's cynical political stage management of the project, and being wary of the glare of publicity that surrounds any human spaceflight activity, let's hold fire for a moment before plunking for Neil Armstrong's adventures as the pinnacle of the first 50 years of the space age.

The thesis of this book is that surveillance satellites had the greatest impact on human history of any form of space activity in the past 50 years, a claim that will be expounded in later chapters. The moon landings are of course a strong candidate for that "50-year" crown, and there are several other contenders for that accolade too. To give you a fuller picture of the historical importance of the first 50

---

[11] Yeager was the first man to fly faster than the speed of sound and forever immortalized in Tom Wolfe's bestseller *The Right Stuff*.

[12] DeGroot (2007) pp. 115–117.

years of spaceflight, let's consider how space has encroached on our lives in that period.

THE FIRST FEW YEARS

Even more than Apollo, Sputnik was a pure weapon of propaganda, containing only a radio transmitter that emitted a shrill beep. It didn't photograph the earth or measure the radiation or temperature of outer space. But Sputnik more than fulfilled the hopes of its creators, winning for the Soviet Union a huge publicity coup. America could no longer claim technological supremacy, a status which had underpinned its political message of the previous decade. America's self-confidence as a nation took a huge knock, while that of the countries of the communist block rose to new heights.

While the propaganda war continued in the form of ever more ambitious human spaceflight projects, it was not long however before the USA and the Soviet Union were putting satellites to use. From the start, using satellites to spy on the other side was the top priority for both superpowers and will be the subject of later chapters. But other uses were soon found.

The first American satellite, Explorer 1, launched just 3 months after Sputnik, measured the conditions of outer space for the first time, thus discovering the Van Allen radiation belts that envelop the globe. The first robotic probes to the moon came soon after.

President Kennedy's famous May 1961 speech announcing the Apollo moon-landing program became a rallying cry for the American space program. But along with his famous commitment "to achieving the goal before this decade is out of landing a man on the moon and returning him safely to earth" was a commitment to develop telecommunications satellites and weather satellites. These other sections of that eloquent call to arms are often forgotten, but they had a dramatic impact on the long-term future of space.

Within 5 years of Sputnik the first commercial telecommunications satellites were being tested, most famously in the form of Telstar which beamed TV signals across the Atlantic in 1962. Telstar's orbit meant that it only linked two distant countries for a short period before it dropped below the horizon, but a year later the first geosynchronous satellite was launched, Syncom-2, demonstrating the concept invented by Arthur C. Clarke in 1945—a satellite at 35,000 km altitude remains above the same point on the earth.

The first successful weather satellite, TIROS, was launched in 1960 by NASA beginning the series of satellites that we now rely on for our weather forecasts.

Indeed, we rely on satellites for many aspects of daily life, often without realizing it. TV is heavily reliant on it. News and sporting events from across the world are carried by satellite. Interviews with commentators, celebrities, and politicians on the other side of town are carried by satellite because it is the cheapest way to relay signals from the outside broadcasting van to the studio.

SATELLITES TODAY—AN OVERVIEW

TV illustrates one of the curious features of space. Space seems expensive and dangerous, but hard-headed businessmen in the media and communications sectors choose to use satellites because they are the most cost-effective option for their business. Likewise for the weather-forecasting agencies, a small fleet of satellites is more effective than a similarly priced suite of ships, buoys, and weather stations. Meanwhile, manned space programs like the International Space Station swallow up billions of dollars, euros, and yen without having much to show for it.

Of the more than 4,000 objects that have been launched beyond the earth's atmosphere in the 50 years since Sputnik, about 600 to 650 are currently working.[13] They fall into four broad categories: scientific, military, commercial, and human spaceflight, so in the pages that follow we will take a quick look at the current state of each of these as the 50 years draws to a close. We tend to call all of these objects "satellites", a term that implies they are in orbit about some celestial body— the earth, the moon, a planet, even the sun. They are also often called spacecraft even when they are totally robotic—by comparison, an "aircraft" is assumed to be manned unless stated otherwise. Some of these objects are on the surface of a celestial body (there are two rovers wandering across Mars as I write this) while others deliberately crash into the celestial body that is their target or head out of the solar system altogether. The term "satellite" should therefore be avoided when referring to these non-orbiting vehicles, and in this book I will call them spacecraft or probes.

First, we might want to consider whether Sputnik-1 and October 4th 1957 were really the start of the space age. The V2 rockets created by Von Braun's team in Germany during World War II were critical in demonstrating the reality of long-distance missile technology. Even though Korolev in the Soviet Union relied to a relatively limited extent on captured German missile technology and personnel, Von Braun's wartime exploits turned the technology from an exotic and very long range vision into a military weapon with immediate and major impact. Military funding for rocket experimentation became much easier to justify after the V2 than before. So, in that sense at least, the rocket that carried Sputnik-1 into orbit owed an important debt of gratitude to Von Braun.

Of course the military impact of the V2 itself in World War II was much more psychological than physical. Indeed, the V2 almost certainly helped the Allies more than the Germans! Historian David Edgerton reports that "the horse made a greater contribution to Nazi conquest than the V2", noting that the Germans required 625,000 horses for the invasion of the Soviet Union in 1941. He also suggests that a missile that flew 300 km to drop a one-ton bomb and then self-destruct was inherently less cost-effective than a bomber aircraft that could drop ten times that over and over again and over greater ranges, and he quotes an estimate of 24,000 fighter planes that could have been built instead of the V2s. In a final irony he notes that the 5,000 victims of the 6,000 V2s were outnumbered by the 10,000 slave laborers

---

[13] Cáceres (2001).

who died in the production of those same V2s.[14] In summary, Edgerton notes that it took two lives to build a V2 and each killed one person—hardly a sustainable business model.[15]

The importance of a technology often depends on when the assessment of its importance is made. Although the V2 actually detracted from the effectiveness of German forces during World War II, it was soon apparent that it represented a technology that had enormous potential, especially with the emergence of nuclear bombs of which only one or two needed to be dropped to have major military impact.

No doubt the purist might wish to go back earlier than the V2 to Robert Goddard in the USA or Tsiolkovsky in the Soviet Union. But Sputnik was unique and special not only for its technical innovation but because it brought spaceflight in from the realms of science fiction.

So, let's have a look at what sort of satellites (or spacecraft) we have created in the 50 years since October 1957—and there have been a lot of them. Just fewer than 4,500 launches, some carrying more than one vehicle, have carried aloft a total of not quite 6,000 spacecraft.[16] As noted earlier, a little over 600 of these are still in orbit and operational.

SCIENCE

Twenty or so of the 600+ spacecraft are way beyond the earth exploring Mars, Venus, Saturn, and other parts of the solar system. The two rovers, Spirit and Opportunity, exploring the surface of Mars are examples of this type of space mission. Space probes are the only way to actually visit the planets, and so our knowledge of them has exploded since space became accessible.

Before the space age Venus was seen as an icon of beauty and love, while Mars was perceived as stormy and warlike. Spacecraft have totally changed our view of these two bodies. Venus, it turns out, is the most hellish of all the planets. Temperatures at its surface average an unbearable 465°C (over 700°F) or twice the temperature inside the family oven. To add to the agony, the atmospheric pressure at the surface is a crushing 90 times that on earth or roughly that 1 km under the ocean. Finally, it rains sulphuric acid, which makes our so-called acid rain seem crystalline pure. The Soviets did manage to land several probes on the surface of Venus but they only lasted a short time before succumbing to the overwhelming temperature and pressure—Venera 13 in 1982 was the longest surviving at just over 2 hours.

Mars, on the other hand, has shown itself to be a land of beautiful scenery, with deep canyons and gorges, tall mountains,[17] and sweeping landscapes. Although seemingly barren of life, it appears to contain large quantities of water below the

---

[14] Another estimate puts the slave labor deaths at more than 20,000: DeGroot (2007) pp. 18–19.

[15] Edgerton (2006) p. 17.

[16] Jung (2007) p. 35.

[17] Giant shield volcanoes rather than the earth's mainly tectonically induced mountains.

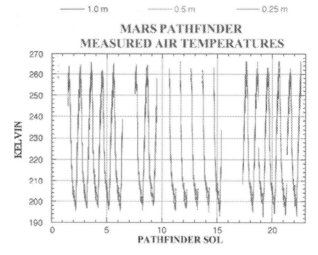

Figure 7. 70°C variation in temperature on Mars between night and day. Courtesy NASA/JPL-Caltech.

surface and captured in polar ice-caps which suggests that primitive forms of life may have existed there in the past—and may continue to do so beneath the intensely irradiated surface. It is cold on Mars, but in summer it's no worse than in the polar regions on earth—unpalatable to humans and our equipment without proper insulation, but within the bounds of our technology (see Figure 7).

Space probes have opened up whole new worlds for us in the outer solar system. The moons around Jupiter and Saturn have shown an enormous range of features that were totally unexpected, with one moon of each of these giant planets apparently capable of supporting primitive life forms. Europa is only the fourth largest of Jupiter's many moons (more than 60 have been discovered to date) but it is covered in ice, unlike the crater-strewn surfaces of all but one of the other moons (Io is the exception, about which more in a moment). Beneath the ice scientists reckon there could be an ocean, perhaps 30 to 50-km deep, giving Europa more water than the earth. If this is the case, organic life may also exist there. The other exceptional Jovian moon, Io, is covered in volcanoes spewing sulfur and other minerals into space. Instead of the tectonic plates that drive earth's volcanoes, Io's are caused by tidal forces induced by Jupiter and the nearby large moons—thousands of times stronger than the tides that influence earth's oceans.

The most recent man-made probe to visit Saturn, the Cassini/Huygens mission, has discovered water ice on the sixth largest of its 50+ moons, Enceladus. Spewed out by geysers probably induced by tidal forces (as for Io above), the ice suggests that liquid water may exist below the surface, and provide an environment suitable to sustain primitive life.

Besides visiting the planets, satellites also give access to parts of the electro-magnetic spectrum that can never be seen from ground-based observatories—earth's atmosphere only lets through visible light, radio radiation, and some forms of infra-

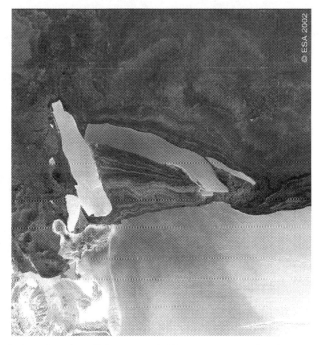

Figure 8. Giant iceberg breaks off the Ross Shelf in July 2002 (mid-winter, so total darkness) monitored by Europe's Envisat SAR sensor. Credit: European Space Agency.

red radiation. The space age has liberated astronomers to explore the universe across the whole of the spectrum. Gamma rays, X-rays, and ultra-violet radiation, for example, are at the high energy end of the spectrum. Violent events in stars and galaxies are visible at these wavelengths, including collisions, explosions, and collapse of these bodies. By contrast radio waves and infra-red wavelengths are associated with the quieter parts of the universe such as the formation of stars from tenuous clouds of dust. Mind you, the expansion of the universe induces a red shift into radiation from distant galaxies, so radiation that may have left such bodies as X-ray or ultra-violet radiation may arrive at earth in the infra-red. For this reason the James Webb Telescope will be designed to observe objects in the infra-red in more detail than is possible with its precursor, the hugely successful Hubble Space Telescope.

Many would say that opening up other planets and indeed the universe to our inspection is the crowning achievement of space in its first 50 years.

Satellites are also telling us new things about the earth itself. From space a satellite can monitor global change, well, globally. Deforestation in the Amazon, shrinkage of the polar ice caps, spreading deserts, and other large-scale phenomena are hard to measure on the ground—it's a case of not seeing the forest for the trees. But space makes these changes clear—see, for example, Figure 8.

About two-thirds of the data used by weather forecasters now comes from satellites, replacing many of the weather ships, instrumented buoys, and weather balloons. Satellites can see the parts of the earth's weather where those older technologies could not reach—especially in the southern hemisphere, which is mostly vast open oceans, but whose weather systems affect us landlubbers days later.

In an era of escalating global warming, the ability of satellites to provide continuous, reliable, and universal data on environmental change will undoubtedly become more and more important.

All in all, about 60 satellites perform scientific research in earth orbit—research into the earth's environment, the sun, and the stars, which together with the 20 or so probes scattered around the solar system make up about 13% of the 600.

MILITARY SATELLITES

About a quarter of the total are military satellites (i.e., 150 or so). America and Russia have the greatest variety and number, with satellites that relay communications, monitor radio traffic, detect radars, ships, and other large military bodies, spot missile launches, provide precision position information, and take images of other countries across the globe. Imaging satellites (the subject of later chapters) may have been the most strategically important during the 1960s and 70s, but these days three other types of satellite are at least as important to the military: satellites for communications, positioning, and radio intercept. Let's take a brief look at each of these.

Military communications

The US military relies almost totally on satellites for communicating with its forces abroad. During the Iraq war in 2003, more than 80% of the communications used by the Allied forces were carried by satellite. The bulk of *that* was provided by commercial satellites, with the military's own communications satellites used for a core of high-security information.

The technology of satellites makes it possible for a war in Asia to be directed from thousands of kilometers away. Planes based hundreds or even thousands of kilometers from the conflict zone are involved. Ships far out at sea provide support in the form of planes, cruise missiles, helicopters, and personnel. Satellite communications make it possible to coordinate these widely scattered forces.

Although the American military has invested billions of dollars in communications satellites, they are sometimes not at the right place at the right time. During the post-9/11 Afghan war, America deployed unmanned aerial vehicles (UAVs) equipped with cameras to provide intelligence of the battlefield. Getting the images back to HQ in the USA proved more than the available military satellites could cope with, so some bright spark decided to send the images by commercial satellite without bothering to encrypt them. A suitable level of commercial encryption is available as a $10 plug-in to your home computer, but the images were sent "in the clear". Browsing through the channels on the commercial satellite an alert South American viewer spotted the images and made them public.

Failing to apply at least some form of encryption to an image from the Afghan conflict is an interesting story, but more importantly it illustrates how central communications satellites are for today's military.

Military forces want communications that are available anywhere in the world at a moment's notice. They want to communicate with ships at sea, with planes and missiles in the air, with troops on the move in tanks and jeeps and on foot. They want to communicate with special forces behind enemy lines without the need for an electrical power generator and a massive radio aerial. They want reliable communications—not the kind you get using short-wave radio or bouncing radio waves off the ionosphere or meteor trails (alternatives to satellites in some situations).

There is a downside to communications satellites in a military context—the satellite is high in the sky and visible to your adversary as well as to your own forces. The adversary can listen in to the radio signals being broadcast to you from the satellite and can aim a strong radio signal at it to swamp its receiver. Military satellites get around these problems by encrypting signals using powerful codes not available to the commercial community. These satellites also conceal the signals by a variety of clever techniques such as hopping from frequency to frequency in a sequence known only to your side. The more sophisticated satellites can change the electronic "shape" of their antennas (the antenna pattern) to shut out a jamming signal while letting your own side's signals through.

The final weapon in the adversary's armory is to physically destroy your satellite. If the adversary has a rocket powerful enough to place a satellite in orbit, he can use it to launch a suicide bomber satellite that homes in on your satellite and explodes when they meet—an anti-satellite (ASAT) missile as these space torpedoes are called. A long-range missile probably can't reach the altitude of most communications satellites (military and civil) which are in geostationary orbit at 35,000 km altitude, although they can knock out surveillance satellites that tend to be in much lower orbits—mostly below 1,000 km. China recently destroyed one of its own satellites in a low orbit in the first overt example of the space torpedo this century.[18] We will return to ASATs in Chapter 10 when we look at what the future may hold.

GPS

Another important use of satellites in the Iraq war was for targeting. Missiles, artillery shells, and bombs almost all now are outfitted with GPS receivers enabling them to steer a precise course to a predefined target. Whereas 10 bombs were required on average to destroy a target in the first Gulf war (1991), that figure was reduced to one in the Iraq war (2003) because of GPS—with a consequential reduction in collateral damage.

Global Positioning System (GPS) satellites are little more than flying clocks. You as a user receive signals from several GPS satellites in the skies overhead, each one telling you its exact location in space at an exact time. The clock in your receiver tells you the time elapsed since the GPS satellite transmitted its signal, which knowing that

[18] Covault (2007a).

the speed of light (and radio waves) is 300,000 km/s, allows your receiver to work out how far away the satellite was. With three of these distance measurements you can work out your position by triangulation. With a fourth satellite you can work out any error in the clock in your receiver, thus improving the accuracy of the distance calculations and that of your resulting position fix.

Twenty-four GPS satellites will ensure that anywhere on the globe at least four of them are within view. The American military usually have about 27 in orbit at any given time—the few extra satellites guarantee that if one fails the system still works.

The information you receive from GPS has been calculated in computers on the ground, and the satellite just broadcasts that information to the users. The clock on a GPS satellite has to be extremely accurate since an error in time of 1 nanosecond (a billionth of a second) equates to a distance error of 30 cm. Clocks with this sort of accuracy have become available only in the past 40 years, going by the general name of atomic clocks. These clocks actually measure frequency—for example, the micro-wave radiation emitted by the cesium isotope with atomic weight 133—but frequency is the inverse of an interval of time, hence their ability to serve as clocks; by counting the wavelengths of the radiation you are counting time.

Measuring the error in clocks with the sort of accuracy you need for precise space navigation was difficult before GPS came along. Part of my job in the Apollo program in the late 1960s was to find ways to analyse errors in the clocks at NASA's tracking stations around the world. Clocks were synchronized to radio signals received from special radio stations in the USA, UK, and elsewhere, but the uncertainty in the time of propagation of those radio signals to the distant station was of the order of milliseconds and certainly nothing like nanoseconds. Other techniques included bouncing signals off the moon to be picked up by the stations, flying a reference clock to each station and then checking its time once it had returned, and simultaneously observing radio emissions from stars or satellites at two widely separated stations.

As was well known to ancient mariners, an error in time was the same as an error in longitude. The first Apollo mission to leave the earth's gravity, Apollo 8 in December 1968, found this out for itself. Tracking data from a third of NASA's stations could not be used because of relatively large errors that were seen when the data was processed. Since Apollo 8 went no closer to the moon than 100 km, the effect was not too serious, but would have been critical for the first landing mission due seven months later. The cause turned out to be poorly determined longitude values for the NASA stations on isolated islands like Hawaii and Ascension. My team at TRW used tracking data from several earlier unmanned lunar missions to calculate the station longitudes and the clock errors at each tracking pass. Fortunately, we were able to distinguish between clock and longitude errors at most of the stations (the Canary Islands site was the main exception) and provide accurate longitude values for Apollo 11—Guam, for example, was in error by over 300 m.[19] On getting out of quarantine in August 1969 after his return from the moon, a grateful Neil

[19] Norris (1994).

Armstrong drove the quarter mile from the Manned Spacecraft Center to our offices to give me the Apollo Individual Achievement Award for this result.

Nowadays, the ultra-accurate clocks on GPS underpin the electronic networks on which we all rely. The digital networks that provide electronic banking, email, the internet, cell phone services, digital TV, and much more all require precise time synchronization to work. One of the reasons the European Union has decided to replicate GPS in the form of its Galileo system is a recognition of the dependence of so much of our economic life on GPS, and a prudent desire to have at least one alternative.

Mariners and explorers use GPS to navigate anywhere in the world. But, even if we never rely on GPS to find our way from A to B, we owe it a vote of thanks for providing the time synchronization that underpins many services that we take for granted.

Electronic eavesdropping

The use of specialized satellites to eavesdrop on radio communications (COMINT) and to analyse radar and other military signals (SIGINT) is still shrouded in secrecy. The world of the imaging satellite has been gradually declassified since the end of the Cold War, but no such opening of the curtains has yet happened in these other forms of spy satellites. There are occasional stories of a mobile phone giving away the location of a wanted person, and bombers then sent in to destroy him. But whether this location was obtained via satellites or via more conventional means on the ground is not made clear.

The American National Security Agency (NSA) is the organization in the USA that unscrambles intercepted communications, employing 35,000 staff. Britain's 6,500-man Government Communications Head Quarters (GCHQ) is its close ally. While these agencies have had to adapt to the explosion of communications in the modern world, they have benefited from the fact that more and more of that traffic is digital, and thus readily analysed by computer.

Emails and text messages are intrinsically computer-compatible, but voice communications still require a degree of sophisticated processing to turn them into a form that a computer can analyse. The strengths and limitations of automatic speech recognition are increasingly apparent as you will know if you have ever encountered a machine when you phone to ask for assistance, get a telephone number, report a fault, or complain. Many such calls result in frustrating misinterpretation by the "intelligent" machine at the other end. The NSA presumably have better algorithms than the phone company for processing voice messages, but the complexity of recognizing any language spoken by any person is daunting.

One small crack in the public awareness of the technical direction followed by these agencies was permitted in 1981 when my company, Logica, was allowed to publicize the award of a contract by GCHQ for a speech recognition machine, able to pick out words in continuous speech. The system worked best if the user "trained" it by speaking a defined vocabulary into it before using it in anger, but it also worked reasonably well on total strangers although it was limited to a vocabulary of about

Figure 9. The world's first spy satellite—the US GRAB ELINT satellite. Credit: Naval Research Laboratory.

2,000 words. It was roughly the size of a small suitcase, and was designed to work in noisy conditions (it used a special algorithm to eliminate noise and other non-words, such as someone clearing their throat). The performance of computer hardware has improved by a factor of more than 100,000 since then, so one can only imagine the capability of today's equivalent machines.

The first ELINT satellite seems to have been the GRAB satellite launched by America less than three years after Sputnik, in June 1960 (see Figure 9). Built and launched in just ten months from President Eisenhower's go-ahead, this tiny satellite weighed less than 20 kg but provided hitherto inaccessible information about radar systems deep inside the Soviet Union. The first successful imaging spy satellite did not return images until August 1960 (see Chapter 4), making GRAB the world's first successful reconnaissance satellite.

GRAB worked by receiving radio signals across a very wide range of frequencies and relaying them to a station on the ground. The satellite didn't process the signals in any way, just acted like a "bent pipe" in taking radio signals that it heard at its 1,000 km altitude and beaming them downwards to a friendly station. By having a station suitably located near the border, signals from 5,500 km inside the Soviet Union could be picked up. Signals from two Soviet radar systems—one associated with SAM-1 anti-aircraft missiles, the other with missile early warning—were detected and could then be analysed. The advantage of knowing the characteristics of an adversary's radar was not only that you could in future know what it was when you detected it, but you could design electronic counter-measures to jam it or confuse it.

GRAB and its successors were able to get an accurate picture of the location and type of air defences across the Soviet Union, information that would have allowed attacking American planes to avoid them.

COMMERCIAL SATELLITES

We've looked at the 13% or so of the world's 600 spacecraft that undertake scientific research, and the 150 or so that pursue military objectives. Almost two-thirds of the 600 in fact are commercial satellites, primarily for communications and broadcasting, with a small number for commercial imaging and weather forecasting.

Since Telstar first beamed TV between continents in 1962, we have become accustomed to having events from across the globe beamed into our homes. I am not thinking of the millions of people who receive their TV signals from a satellite, but about the stories beamed to the TV studio from around the world. Getting the TV signal from the studio to your home is usually by conventional radio, or cable or sometimes satellite, and nowadays increasingly via the internet. But the real magic is how the stories get to the studio in the first place.

Before the days of satellites, instant news was only available by radio. I remember as a boy listening to the crackly broadcast of the Melbourne Olympic Games in 1956. It seemed exciting at the time, but we had to wait several days before images were available, flown from Australia onboard commercial airlines. By the time of the 1964 Games in Tokyo, Syncom-2 was in orbit and able to beam the TV pictures to America. Since then, we take it for granted that newsworthy events no matter in which part of the world will be beamed to us instantly. Wars, famines, earthquakes, forest fires, floods, in fact any disaster is brought to us in graphic detail via satellite. Who can forget the images of an American cruise missile whistling past a hotel room window in Baghdad, or a student standing in front of a tank in Tiananmen Square?

And this instant reporting of world events affects the way politicians respond. Logical, prudent policies are often brushed aside in the need to answer the call for action by a clamoring media backed up by the latest images from the crisis zone. The Vietnam War was perhaps the first conflict influenced in this way. TV pictures of American soldiers being killed in action or of other horrors of that war helped create a climate of opinion that eventually forced America to withdraw.

On a lighter note, satellites beam TV pictures of sporting events across the globe. The Olympic Games and the soccer World Cup are the two largest sports extravaganzas that now rely on satellites to beam events to every country in the world. The cultural and social changes wrought by this globalization of sport is surely significant, but has perhaps not been thoroughly studied.

The cultural impact of TV was perhaps most strongly demonstrated in its Westernizing influence on eastern Europe in the 1980s. Homes in the communist countries of eastern Europe close to the West could pick up Western TV, which differed from the drab state-run propaganda on their own channels. Seeing the better quality of life in the West on TV, the trickle of refugees leaving eastern Europe for the West swelled to a stream and then a torrent, until eventually the governments in the East collapsed.

Both terrestrial and satellite TV were involved in ending the Cold War, but today we are witnessing a more subtle cultural transformation triggered specifically by satellite TV. The country this time is India, and since the government relaxed the rules about foreign TV broadcasts, viewers have begun to receive Western TV programs via satellite. Up till now Indian programs have not shown men and women kissing, and have respected other long-standing social rules. Western TV does not constrain itself in this way, and the result is a subtle and gradual Westernization of Indian culture.

For a decade or so after communications satellites became available, one of their most important functions was to provide affordable and high-quality intercontinental telephone connections. Before 1965 a trans-Atlantic phone call had to be booked in advance and used radio links to provide a poor quality line prone to hissing, distortions, fading, and general noise. And it cost half an average man's weekly pay packet. By the end of the 1960s the quality had risen and the price reduced to that of a long-distance call. The explosion in communications opened up by satellites (for both TV and telephone calls) proved commercially attractive but technological advances in the undersea cable industry soon challenged the role of satellites, particularly when optical fiber technology became available—the first trans-Atlantic example being TAT-8 in 1988.

By the 1990s undersea cable was carrying the bulk of communications between the world's main population centers, and many people predicted that communications satellites would become a thing of the past. These pessimists were proved wrong and the communications satellite industry continued to expand. More recently the pessimists forecast a reduction in communications satellites because of the move from analog to digital TV—digital channels require less bandwidth than analog, so fewer satellites are needed to carry the same number of channels, went the argument. In fact, the number and size of satellites continues to increase.

Communications satellites have prospered because they fill a number of niches very cost-effectively. And as an old niche disappears, as long-distance telephony did in the 1980s and 1990s, a new one emerges. Let's look at some of today's niches.

If you want to talk or send email on the move then you either use a cell phone network or satellite—cable doesn't follow you to your car. Inmarsat satellite terminals that link to the specialist satellites addressing this market are now universal on ships of any size and on most large aircraft. They have even become a fashion statement for rich adventurers—in the movie *Blood Diamonds*, Leonardo DiCaprio depends on an Iridium handset to orchestrate key events in the story such as calling an air strike on a terrorist camp, and even the tear-jerking final farewell to Jennifer Connolly is by satphone (as *he* calls it).

DiCaprio borrowed or stole (I lost the plot a bit at that point) his satphone from an aid agency. And emergency agencies do rely heavily on communications satellites. Typically, their work is in an area without modern infrastructure, or where that infrastructure has been damaged. In those circumstances satellites are the communications bearer of choice.

In fact, a specialist aid agency has been created to provide communications satellite services to crisis areas. Télécoms Sans Frontières (TSF) like its more famous near namesake Médecins Sans Frontières is based in France. Founded in 1998, it has helped survivors and agencies in more than 60 countries. In 2006 the UN signed its first ever worldwide agreement with a non-governmental organization (NGO)—a "First Responder" agreement with TSF. "There is an urgent need for food, water, shelter, protection and medical help in emergencies," said Ann M. Veneman, Executive Director of UNICEF. "None of these things are possible without quick and reliable communications. Rapid communications saves lives." The founders of TSF Jean-François Cazenave, its President, and Monique Lanne-Petit, its Director, noted that "Access to reliable communications in the first hours following a crisis strengthens the coordination of agencies that save lives and respond to the needs of victims".

Mr. Cazenave recalls that when he used to work for conventional aid agencies in crisis zones, victims would seek him out and press a piece of paper into his hand with a telephone number and name written on it asking him to call that number when he returned to civilization. He noticed the high priority that many victims gave to establishing contact with family members, to report that they were safe or to find out about other family members. This observation led him to found his organization initially with a single Inmarsat satellite phone. Now TSF operates with 13 permanent staff and about 40 volunteers from three centers in France, Nicaragua, and Thailand. TSF depends on the charity of organizations like Inmarsat and Iridium to donate satellite time free—as you might expect, satphone charges are more expensive than a cell phone, which raises the question of who was picking up the charges for Mr. DiCaprio's calls in *Blood Diamonds*.

In December 2006, TSF was in the Indonesian island of Catanduanes (300 km east of Djakarta), helping the survivors of Typhoon Durian. They enabled over 800 families to make contact with loved ones (see Figure 10) of which over 200 called abroad, to the USA, Canada, the UK, Hong Kong, Australia, Kuwait, the Emirates, Germany, etc. often to give news but also to ask for help and especially money.

I said that long-distance telephony had disappeared as a niche for communications satellites, but in fact it continues to prosper for those countries not lucky enough to be connected to a long-distance cable system. For many island nations in the Pacific and Indian Oceans and the Caribbean, satellite communications are the only reliable connection to the outside world. The same is true for many developing countries in Africa and South America. Initially, the social importance of communications satellites was so great that the Intelsat and Inmarsat organizations were established by international treaty to manage that business—Inmarsat to mobile terminals, Intelsat to so-called fixed terminals. With the proliferation of undersea cable and of commercial satellites both organizations have now been privatized and operate as commercial companies—with one exception: both organizations are still required to provide global services on a fair and equitable basis to every country.

Figure 10. Télécoms Sans Frontières in action in the Philippines in 2006. Credit: Télécoms Sans Frontières.

I said that new niches keep appearing for communications satellites. In the USA recently one of the latest such niches is high-quality radio services offered by two competing satellite companies: XM-Radio and SIRIUS, which together now serve more than 13 million subscribers across North America.

Another emerging niche is high-definition television. Traditional broadcasters generally lack the radio spectrum to transmit the spectrum-hogging high-definition form of TV, and not all cable companies have the necessary bandwidth either. Satellites have been able to jump in and offer nationwide HDTV services without the need to install new cables or relay stations.

The ability to deploy a new service instantaneously across a whole country or continent has proved an important advantage of satellite. In Britain the company awarded the licence to run the national lottery game had only six months from licence award in which to deploy thousands of ticket machines and connect them reliably to headquarters. They chose to use satellite dishes at the premises of each ticket machine as the fastest and surest way to be up and running in time—about 4,500 satellite terminals had been installed by the time they went live.

Another example of rapid deployment is the use by TV journalists of satellite technology for filing their reports from the field. You can understand that a satphone is the only realistic option for a reporter in a war zone or after an earthquake or flood, and it is interesting to note the immediate increase in quality of satphone reports from Iraq on TV once Inmarsat launched its new Inmarsat-4 broadband satellite in 2005. But satellites are used by outside broadcast units just the other side of town from the studio, or from outside the White House where one assumes that a conventional communications infrastructure is readily available. It turns out that the networks like the flexibility of satellite—OK it's not essential at the White House, but it will be to follow a story of a shoot-out or traffic accident elsewhere in DC. As mobile broadband technologies like WiMax and 3-G proliferate, it will be interesting to see if communications satellites retain this niche other than just in crisis areas and in the countryside.

Overall, I find that the case for communications satellites as the flag bearer of the space age's first 50 years is strong—globalizing our view of the world, opening the eyes of millions to events on the other side of the globe, forcing politicians to respond to events as they unfold, saving thousands of lives at sea and in crisis zones, and addressing the desires and needs of millions across the world. Not all of these effects are necessarily good (the instant political response triggered by satellite reporting may lead to unduly hasty decisions, for example), but they affect us all.

HUMAN SPACEFLIGHT

We started this chapter by recognizing the iconic impact of Apollo 11 when Neil Armstrong stepped onto the surface of the moon in July 1969. And if a single event defines the most important contribution of satellites in the 50 years since Sputnik, then Armstrong's one small step gets most people's vote.

However, the *last* man to step on the moon's surface did so quite soon after Armstrong—Gene Cernan when he re-entered the Apollo 17 Lunar Lander in the Taurus Littrow Hills on December 15th 1972—more than a third of a century ago. Landing a man on the moon was a great technical achievement but seems to have been something of a dead end, at least for the moment. Every five or ten years since then the US government issues grand plans to return humans to the moon, and these plans then slowly (sometimes rapidly) wither in the cold clear light of funding priorities. Now, on the 50th anniversary of Sputnik-1 we are in the early stages of one such "back to the moon" cycle. The declarations are made, the budget estimates begin to rise, the scheduled date begins to slip, and the other programs begin to consume the available funds—thus far no change. Next will be the call for international partners to join the great endeavor, the inevitable schedule slips while the partners negotiate their roles, and so on. Perhaps this time will be different, but so far the picture of the history of the 1980s and 1990s repeating itself seems strong.

"Dead end!" I hear you exclaim. "What about all the incidental benefits we got from the Apollo missions? What about the invention of the Teflon frying pan and the body scanner?" Ah yes, the "spin-off" benefits. We should indeed take a look at them.

The Teflon frying pan is often suggested as a useful fringe outcome of the Apollo missions—NASA was quoted in the *Wall Street Journal* in 1969 claiming it as a "major example of space spin-off".[20] Perhaps the articulators of this argument are fooled by a similarity between the technology of the non-stick surface and that of the re-entry shield on the nose cone of the returning Apollo capsule. The trouble is that Teflon was discovered by the Du Pont company in 1938, and used during World War II mainly in the making of bombs (it gets a mention in that regard in Chapter 3). The use of Teflon in non-stick frying pans was introduced in 1954 by a Frenchman Marc Grégoire under the brand name Tefal (TEFlon + ALuminum). By the time President Kennedy kick-started Apollo in 1961, Tefal was selling a million frying pans a month in the USA.[21]

The body scanner is another Apollo spin-off myth. Invented by EMI in the UK in the early 1970s, they sold off their interest in the technology and closed the plant near London about 10 years later, whereupon many of the technical staff came to work with me at software company Logica. We have discussed this myth over the years and I can vouch for the fact that the origins of the technology lay more in military radar developments than in any space program.

By and large, the story of human spaceflight since the days of Apollo is one of a technology seeking an objective. The Soviet Union/Russia and the USA (and now China) have ensured that astronauts have been in space nearly continuously for the past three decades. And those activities have been widely publicized by the mass media. Part of the attraction is of course the danger. The two horrific Space Shuttle disasters have taken place in the full glare of a press frenzy. For the next few Shuttle launches after a disaster, the world holds its breath in fearful anticipation. Then, after five or six uneventful flights, press attention wanes, and with it the willingness of Congress to fund the program.

The sad fact is that little real benefit has come from the human spaceflight of the past 30 years, in comparison with its cost. The processing of materials in the absence of gravity is an interesting area of research, but the last thing you want when doing such research is astronauts walking around the spacecraft creating tiny disturbances. Far better and cheaper to run such experiments on a robotic satellite with a radio link to earth for direct control of the equipment if necessary—tele-operated, as they say.

Astronauts can fix satellites in orbit—the Hubble Space Telescope has been repaired and enhanced on a number of occasions by a visiting Space Shuttle crew. But at what cost? NASA usually gives the cost of a Shuttle mission as about $500 million, but that's only part of the story. The Shuttle accounts for typically $4 billion out of NASA's annual budget, and if they are lucky that gets them five Shuttle flights—recently, it has been averaging about one flight per year. Whatever way you do the maths the cost comes to close to or above a cool billion dollars per Shuttle launch. Worse still, as the Shuttles get older they get more and more expensive to keep in working order, so the costs will rise. And we still have to add in the cost of the equipment and material it carries. Equipment that flies on the Shuttle has to be

---

[20] DeGroot (2007) p. 149.
[21] Edgerton (2006) p. 19.

designed and tested to a higher level of safety than if it is launched on a normal rocket. How much extra this safety proofing costs is difficult to measure, but it's a lot. So adding in the cost of the safety-proofed parts used to repair the Hubble Space Telescope, the cost of a Shuttle repair flight is well over $1 billion and probably closer to double that. Even at the lower end of that cost estimate, you could build a brand new and better version of Hubble and launch it on a conventional rocket, because it wouldn't have to be safety-proofed.

Aviation inventor Burt Rutan has been asking NASA for the past 30 years when he could buy a ride into space at a not-astronomical price. The answer he consistently received was "30 years". After 20 years of getting that same answer Burt decided if NASA wasn't going to do it, he would have to do it himself. And he's got part way there—as far as Al Shepard did in Mercury in 1961. With successive launches of SpaceShipOne 100 km high in 2003—a so-called suborbital flight—he claimed the $10 million X-Prize awarded by entrepreneur Peter Diamandis and his backers. Now Burt is building the SpaceShipTwo commercial variants of that vehicle in which paying passengers will be able to visit space and experience several minutes of weightless flight while viewing the curved earth below. He is also determined to reach the next step of getting into orbit not just the up and down ride of SpaceShipTwo. The airline billionaire Richard Branson, who is funding the commercial operation of Burt's vehicles, is also keen to take the initiative all the way to orbit, and if SpaceShipTwo is commercially successful he seems likely to back further development. Burt's inventiveness and aviation expertise has been instrumental in creating the only serious alternative to the government-sponsored rockets so far on the market, so it will be interesting to see if he can break human spaceflight away from NASA's wildly expensive technology.

The fact is that robotic technology has improved a million times over since the days of Apollo—the best known example is Moore's law for computer chips which for 40 years has held that the number of transistors on a chip doubles every 18 months or so. A one-ton satellite is therefore nearly a million times more powerful than it was 40 years ago, but a human in space still weighs the same, consumes the same amount of food, air, and water, requires the same volume of habitable space, and is no stronger or cleverer than before. It becomes more not less difficult therefore to justify sending astronauts into space for any reason other than national prestige or inspiration of youth. Meanwhile, millions can savor the beauty of Mars or Saturn via high-quality TV images from robotic vehicles.

The events of the past 50 years suggest that human spaceflight is only one of four space stories that captures the attention of the general public and the mass media. My list of the four headline-grabbing space topics is as follows:

1. *The threat of an asteroid hitting the earth.* Since the Alvarez father and son team postulated in 1980 that the extinction of the dinosaurs was caused by a cataclysmic asteroid impact, this story has held the public's attention, and nature conveniently set up a demonstration of the event by arranging for the comet Shoemaker–Levy 9 to hit Jupiter in 1994. The small comet or meteor that exploded over the skies of Tunguska in the depths of Siberia in 1908 with a

force equivalent to a 10–20 megaton hydrogen bomb is the most recent major example of what few now doubt is a serious long-term threat to mankind.

2. *Discovery of life beyond the earth.* Scientific opinion waxes and wanes on whether at the one extreme the universe is populated with intelligent beings or at the other extreme humans are unique.[22] Discovery of intelligent aliens would of course be a show-stopping event, but what seems more likely in the near term is for primitive life forms to be found. Mars, Jupiter's moon Europa, Saturn's moon Enceladus, the atmospheres of Jupiter or Venus are all considered potential locations to find microbe-sized life forms. We find these hardy living creatures in the most hostile places on earth so we can't yet exclude the possibility that they occur elsewhere.

3. *Landing robotic vehicles on Mars.* The scenery of Mars is so stunning, and the chances of finding primitive life forms so tantalizingly possible, that even un-manned robotic vehicles make the headlines. NASA's Spirit and Opportunity are the most dramatic examples to date, but the strength of this topic was perhaps most strikingly demonstrated by Britain's shoestring attempt to place a lander on Mars—Beagle 2. For three years—as the project collapsed, rose again, wobbled, and steadied—its charismatic leader Colin Pillinger was a constant feature on TV and in the newspapers. The vehicle was last seen leaving its mother-ship, Mars Express, scheduled to land four days later on Christmas Day 2003, but alas it was never heard from again. For a space mission to capture the public imagination in space-phobic Britain (the only G8 country not involved in human spaceflight) demonstrates the drawing power of this topic.

4. *Human spaceflight.* The electric impact of astronauts and cosmonauts is still strong. They attract crowds—politicians, journalists, and the general public. They are particularly popular with young people, projecting an image of lofty technical superiority and bravery. Ticker-tape parades down Fifth Avenue in New York were the norm for the early astronauts, and although that level of public adoration is no longer prevalent, their aura is still strong. The first Chinese taikonaut and the first Swedish astronaut have demonstrated in the past four years the degree of national frenzy these apparent super-heroes continue to cause.

Looking ahead, I offer one more headline-grabbing space story of the future to add to the four above:

5. Discovery of an earth-like planet in another solar system. In the past ten years astronomers have begun to detect planets around other stars—more than 200 to date. In most cases, "detect" means spotting tiny wobbles in the star's orbit caused by the gravity field of the planet or tiny variations in the light of the star as the planet crosses its face. Within the next 10 or 15 years it seems likely that telescopes in orbit or perhaps here on earth will be improved to the point where

[22] See, for example, Ward and Brownlee (1999) and Webb (2002).

these faraway planets can be imaged, and methane or ozone detected in their atmospheres, good indicators for earth-like conditions.

You will note that surveillance satellites are not on the list of headline-grabbing space accomplishments—the general public has not shown the same emotional interest in them as in astronauts, asteroids, extraterrestrial life, and the other types of space mission listed above. The sharp distinction between the intrinsic importance of spy satellites and the lack of public endorsement of their value is nothing new. In the 1950s and 1960s several commentators in the know summarized the difference between the glamor of human spaceflight and the harsh functional realism of surveillance satellites with the observation that "one steadied the resolve of the American public; the other steadied the resolve of the American President".[23] In the remainder of this book we will explore how *the President's resolve* was steadied.

[23] See, for example, Albert D. Wheelon, in Day *et al.* (1998) p. 38.

3

Cold War nuclear stand-off

The first ballistic missiles were the V2 rockets that rained down on London and the southeast of England in 1944 and 1945. The brain-child of German engineer Werner Von Braun, the V2 was amazing in that it demonstrated how rocket power could propel a bomb hundreds of kilometers. Previously, the range of an artillery shell was limited to a few kilometers, with a few monster guns able to achieve a range of perhaps 30 km.

The psychological effect of the V2 was massive, but as discussed in the last chapter its military effectiveness was minor—negative even. It was almost an unguided missile in that it landed only very generally in the area of the pre-determined target. A target as big as London was relatively easy to hit—somewhere. But military targets such as an airfield or a garrison were too small for the V2 to hit. It was a miracle of engineering that the V2 flew at all, never mind reaching a target with precision. It was exciting engineering, but more a psychological weapon than one that could damage the enemy physically.

Hiroshima and Nagasaki changed this. On August 6th and 10th 1945 each of these Japanese cities was obliterated by a single bomb—Figures 11 and 12. The atomic bombs as they became known had the destructive power of about 15,000 tons of TNT, or a 15-kiloton yield to use the terminology that the media quickly picked up. You didn't need a precision delivery to take out an airfield if everything within 5 km was flattened, a missile that landed within a few kilometers would do the job.

There would of course be collateral damage in any such military use of atomic bombs. The term *collateral damage* is a relatively recent invention, but the massive death toll of civilians associated with just a single atomic bomb was widely known and discussed. Massed World War II bombing raids had killed tens of thousands in a single night in Berlin, Tokyo, and Dresden, but those raids had required hundreds of bomber aircraft. Hiroshima and Nagasaki achieved death tolls of more than 100,000 with just a single aircraft and a single bomb—this kind of death toll had only been

Figure 11. August 6th 1945, the Japanese city of Hiroshima is destroyed by a 15-kiloton atom. bomb

achieved once before: in the firestorm that swept Tokyo on March 10th that same year created by the incendiary bombs from 325 US B-29 bombers. The magnitude of the death tolls is a matter of debate still, if only because many people died later from radiation poisoning or from cancer triggered by the blasts. Including those long-term casualties the Hiroshima death toll was about 200,000 and that of Nagasaki about 140,000. The lower Nagasaki total was probably because the bomb was dropped not in the center of the city but blindly through thick cloud despite orders to ensure clear visibility of the target. The pilot's eagerness to get rid of the bomb was because he realized he hadn't enough fuel to get home unless he did so—the shortage of fuel was caused by a faulty fuel pump and by the fact that the plane had diverted from its primary target (the Kokura Arsenal 150 km to the north) having encountered strong Japanese air defences there.[24]

The stakes were raised yet again in 1952 when the first hydrogen bomb was exploded by the USA on a test range on the Pacific island atoll of Eniwetok, with a yield in excess of 10 million tons of TNT (i.e., 10 megatons). I will have more to say about the H-bomb in Chapter 6.

[24] DeGroot (2005) pp. 95–101.

Figure 12. August 10th 1945, the second atom bomb dropped on Japan destroys the city of Nagasaki.

This chapter takes a look at the paranoia of the political leaders on both sides and how close that brought the world to the brink of annihilation. To begin, we consider the source of the danger facing the world—the bomb and the missiles that could deliver it.

THE BOMB

The atomic bomb works because a uranium atom is the largest of all of the 92 naturally occurring elements, and its size makes it inherently unstable as shown by the fact that it emits radiation (i.e., it is radioactive). The radioactivity happens because uranium atoms gradually split apart into an element a bit smaller than itself such as lead and into ionized helium (the second smallest of the elements); the ionized helium (called an alpha particle) is ejected at speed and is the radioactivity we can measure on a Geiger counter. What's left behind is the lead (i.e., the uranium *decays* into lead). Not all at once, mind you; it takes billions of years for the main form of uranium to decay.

The nucleus or heart of an atom is made up of a mix of protons and neutrons. The number of protons determines what the element is—8 protons for oxygen, 26 for iron, etc. The number of neutrons in a nucleus is usually a bit more than the number of protons but can vary a bit and doesn't change the element. Iron with 56 neutrons and iron with 54 neutrons are both iron—they are different *isotopes* of iron, but look, taste, feel, act exactly alike (well, very nearly exactly as we shall see below). Usually, only a few proton/neutron combinations of an atom are stable, the others decaying rapidly through radioactivity if they exist at all. One of the best known examples is carbon 14, a radioactive isotope of carbon that can be used to date organic material dating back tens of thousands of years. Another topical example is polonium 210 which seems to be the current poison of choice for eliminating Russian dissidents abroad (gunshot seems to work best at home). Like uranium, when it decays polonium 210 radiates alpha particles, which can be stopped with a sheet of paper. So it's not dangerous then? Well, yes, it is because if instead of a sheet of paper you use the organs of your body to stop the alpha particles they become badly damaged and eventually fail. Polonium 210 consumed in food or drink therefore ends up giving you massive internal organ failures. It's deadly to human tissue but quite safe to carry around in a cardboard box.

When uranium splits, in addition to the lead and the alpha particles, it gives off some energy. If all the atoms of a chunk of uranium split simultaneously you get an atomic bomb, but fortunately in nature they split one at a time, taking a long time (billions of years) to completely morph into lead, which is why incidentally they are still found on earth four and a half billion years after its formation.

Uranium has 92 protons in its nucleus. All of its isotopes are radioactive, some more than others. The isotope with 143 neutrons turns out to be particularly interesting—it is called U^{235} where 235 is the number of protons plus neutrons. Instead of waiting for U^{235} to decay, if you bombard it with neutrons it will fall apart into two nearly equally sized elements plus a few extra neutrons. So, you put in one neutron to U^{235} and get out some miscellaneous elements plus more neutrons than you put in. If those extra neutrons can then split open more U^{235} atoms that gives rise to even more neutrons, and so on *ad infinitum*—what's called a *chain reaction*. Each time one of the uranium atom splits it also gives off energy. If the chunk of U^{235} is large enough, the encouragement to split grows exponentially and all of the U^{235} atoms (there are about two and a half trillion trillion of them in a kg of uranium[25]) split with a gigantic release of energy.

The splitting of an atom is called *fission*, so the atomic bomb is also called the fission bomb.

All sounds pretty straightforward, but there are a few catches. First of all, U^{235} makes up only a tiny part of the uranium found naturally on earth—0.7% to be exact. The other 99.3% is U^{238} which is also radioactive but is not sufficiently unstable to generate a runaway chain reaction. It's hard to separate U^{235} from all the U^{238} because they are pretty much identical. The main difference between them is that one is slightly heavier than the other, and this can be used to separate the U^{235} from the U^{238}—but with difficulty. First, you chemically combine uranium with fluorine (nasty stuff, used in mustard gas in World War I) to give you uranium hexafluoride,

---

[25] A trillion is a million million.

which is usually shortened to *hex*. Nowadays, you spin the hex in a centrifuge a bit like a giant version of the ones they use on CSI shows on TV. Eventually, the heavier U^{238} has to some extent been pushed to the outside of the spinning tube and the gas on the inside has more U^{235} in it—not much more, but a bit. You take the inner gas that is slightly enriched with U^{235} and put it into another centrifuge where it gets enriched a bit more. Eventually, after a lot of these centrifuges (hundreds) the U^{235} has been enriched from its initial 0.7% to 70–80%, which is good enough for an atom bomb—weapons grade, as they say. By the way, hex is extremely corrosive attacking pretty much any metal, except nickel and special types of stainless steel.

During World War II the centrifuges weren't up to this task—the materials out of which they were made wouldn't allow them to spin at the required high speeds. Very quickly therefore the US and Britain switched to a method called gaseous diffusion where the hex is pushed through a fine mesh grid with the result that the U^{235} goes through a bit faster than the U^{238}—not much faster, but a bit. Push it through enough fine mesh grids and eventually the proportion of U^{235} will be good enough. It took a while and a lot of money before the US found a fine mesh grid that would separate the two uranium isotopes and not get fouled up in the process—a compressed powder of nickel.[26] The now ubiquitous Teflon was used to coat any seals and valves, and then after the war for pots and pans.[27] The Soviets spent less money and time discovering that centrifuges weren't up to it and that gaseous diffusion was the best enrichment technique—their spies told them what the US had discovered in this regard.

The most important of the Soviet spies was Klaus Fuchs, a German refugee who came to Britain in 1933 and was transferred to the US to assist with the Manhattan Project to develop the atom bomb. In 1946 he returned to Britain with wide knowledge of the design decisions taken for the American bomb and even for the proposed H-bomb. Throughout his time in both Britain and the US, Fuchs fed reams of information to the Soviet Union where it was eagerly read, pointing them, for example, at gaseous diffusion instead of the centrifuges they were planning to use for uranium enrichment.

Centrifuges are nowadays the commonest technology used for enriching uranium. Pakistan stole centrifuge plans from the Netherlands to kick-start its nuclear program, and Iran is in dispute with the UN about whether it is enriching its uranium too much with its centrifuges—enrichment to about 8% is a good idea for a civil nuclear reactor, but anything above 20% is clearly aimed at making a bomb.

Incidentally, people usually say that isotopes of an element are chemically identical, but that's not strictly true. The commonest way to separate the various isotopes of hydrogen is to take advantage of the fact that the lighter isotopes evaporate faster than the heavier ones (i.e., using a chemical process). This process doesn't work for uranium simply because hex^{235} and hex^{238} are too close in weight just under 1% difference. But the three isotopes of hydrogen differ by 100% and 50%, making the evaporation technique practical. New techniques involving magnetic separation of hex ions created by laser excitation have been tested in the laboratory

[26] Rhodes (1995) p. 92.
[27] DeGroot (2005) p. 49.

and give concern that they might be more affordable than the dauntingly expensive centrifuge and other methods, thus facilitating proliferation of this dangerous process.

Although U^{238} doesn't make good weapon fuel it works fine in a nuclear reactor—to generate electricity, for example. The U^{238} in a reactor absorbs a neutron and becomes a new element called plutonium, specifically the 239 isotope of plutonium, Pu^{239}. It turns out that Pu^{239} works fine in a bomb, in fact it's even more explosive than U^{235} in the sense that a smaller quantity will explode—the Nagasaki bomb contained just 6.2 kg of plutonium compared with the 55 kg of uranium in the Hiroshima bomb. Plutonium is one of those elements that doesn't occur naturally on earth—any that was here when the earth formed has decayed by radioactivity.

The big advantage of plutonium is that it *is* chemically different from uranium, so it can be separated from the uranium much more easily than separating two isotopes of uranium from each other. So, you start up your nuclear reactor using ordinary uranium and after a while you remove the uranium fuel and chemically separate the newly created plutonium from it. Bomb historian Richard Rhodes reckons that this was the single most important atomic bomb secret the Soviets picked up from their spies, saving themselves at least two years.[28] The critical importance of plutonium for the Soviets was that they had access to very little uranium—not like the large quantities of Belgian Congo uranium ore that the US was getting. The difficulty with a U^{235} bomb was that you needed a lot of uranium out of which to extract the tiny proportion of it that was U^{235}.

Mind you, plutonium is an exotic material and the Americans spent quite a lot of the early years learning about it, but Klaus Fuchs helped the Soviets bypass some of the long learning curve. For example, plutonium has five phase transitions between room temperature and its melting point, some causing it to contract, others to expand. The densest phase was the preferred one since it minimized the critical mass, and Fuchs advised the Soviets that this otherwise volatile phase could be stabilized by alloying plutonium with the rare metal gallium. Fuchs also described in detail the design of the first US bomb—the one exploded in the Nevada desert on July 16th 1945, which is why the first Soviet bomb looks like its carbon copy.[29] We will return to Fuchs's contribution to the Soviet H-bomb work in Chapter 6.

The Soviets bizarrely even got atomic bomb material from the US under Lend Lease—an arrangement for the US to provide war materiel to the Soviet Union to help in the fight against Nazi Germany. Most of the $11.5 billion (these are 1945 dollar billions, remember) went for ships and guns and planes and bombs. But, apparently, the Soviet Lend Lease representatives in the US would order raw materials needed for the Soviet nuclear program. This aid to an ally fighting almost alone against Germany from 1941 until mid-1944 had very high priority and over-rode almost any other consideration. The list of goods relevant to building a nuclear reactor included beryllium, cobalt, heavy water, and even a small quantity of uranium—the US personnel dealing with that last item claim to have ensured that

[28] Rhodes (1995) pp. 74–75.
[29] Rhodes (1995) p. 192.

the uranium was very low quality. The planes taking Lend Lease to the Soviet Union brought people in the other direction—spies, presumably.[30]

The atomic bomb works because of the energy left over when you *split* the uranium atom. Scientists realized that there is also energy left over when you *combine* two atoms into a single new atom. This process of *fusion* is what powers the sun, indirectly therefore providing the light and heat which ensures our survival. Hydrogen, the smallest of the elements, is the most efficient element in this respect, and when atoms of hydrogen combine or fuse to form the next element up, helium, they release a relatively large amount of energy—more than released when an atom of U^{238} splits. The catch is that atoms of hydrogen will only fuse when compressed together at extreme pressures. If enough compressed hydrogen atoms are nearby the energy generated by the first fused atoms will start a chain reaction that goes on until all the hydrogen is converted into helium. The 15-kiloton Hiroshima blast was obtained with 55 kg of U^{235} but the same explosive power could in principle be obtained with a fusion bomb containing less than 200 g of hydrogen. The relatively rare isotope of hydrogen containing two neutrons—deuterium—and the even rarer one with three neutrons—tritium—fuse even more effectively than the common single-neutron hydrogen isotope.

The pressures and temperatures at the center of the sun and other stars are sufficient to force hydrogen atoms together. But the only place on earth with the same conditions as the center of the sun is at the heart of an atomic bomb, and American bomb designers very quickly realized this. By November 1952 they had manufactured an atomic bomb with a fusion bomb at its center, so when the atomic bomb exploded it set off the even more powerful hydrogen bomb.

America didn't keep the secret of either the atomic or hydrogen bombs very long. By a mix of inventiveness and spying the Soviets soon had both forms of nuclear bomb—the atom bomb by August 1949 four years after the US, and the H-bomb by August 1953 only ten months after the US. Britain's first atom bomb was tested in March 1952—the closeness of timing and the similarity of design of all three countries largely the result of Klaus Fuchs—other UK scientists who worked on the wartime atomic bomb project at Los Alamos also brought back useful information for the UK bomb effort—chief among them William Penney who headed the British project.

It may seem obvious now that America continued to develop atomic bombs after the War, but it was hotly debated at the time. Many pointed out that it was not that cost-effective, requiring an enormous industrial infrastructure to enrich the fuel and manufacture the bomb. Others noted that its horrific collateral damage made it impractical to use in anything other than a "war to the death" situation. The debate hotted up even further when the subject of the H-bomb was introduced. Surely, a bomb that could destroy a city like Hiroshima was sufficient? What conceivable target could you have for which the atomic bomb would not suffice? In the end the decision to continue producing atomic bombs and to develop the H-bomb was taken because it was feared the Soviets would do so—and the Soviets did it because they feared the US would do it. The recommendation to President Truman from the

[30] Rhodes (1995) pp. 94–101.

Joint Chiefs of Staff to develop the H-bomb said "the United States would be in an intolerable position if a possible enemy possessed the bomb and the United States did not".

Thanks to Fuchs, the Soviets had seen the H-bomb design drawn up by the Los Alamos team during the war, and they assumed that the US would be pressing ahead with its development. But, in fact, there was a four and a half year hiatus after the War before the US did indeed decide to press ahead. There was initially so little support for an H-bomb that the first time President Truman ever heard of it was October 1949, and he only gave the go-ahead for its development the following January. When his advisers came to him to argue the case for proceeding with the development, expecting a difficult debate, he said "what are we waiting for?"—he had bought the argument of the Joint Chiefs—if the Soviets are going to have it, we'd better have it first.[31]

Ironically, the H-bomb design that the Soviets had seen (Teller's *alarm clock* device) was a dead end, and one that the US rejected, changing in early 1951 to the eventually successful Teller–Ulam "two stage" design—see chapter 6. As a result, the first Soviet H-bomb test produced a much smaller explosion than expected.

DELIVERING THE BOMB

Both America and the Soviets also had the missile technology that Germany had demonstrated with the V-2 in 1944–1945. Von Braun and most of his engineering team moved to America and willingly shared their know-how with their new hosts. The Soviets occupied many of the German V2 manufacturing facilities, but captured only a handful of the key engineers. Nevertheless, in their own Sergei Korolev, the Soviets had a designer and organizer of the caliber and charisma of Von Braun, and he soon developed Soviet missile technology that surpassed anything the USA even with Von Braun's help was achieving.

By 1957 both America and the Soviets had missiles that could carry a nuclear warhead across the globe. The maximum range of such missiles gradually increased: first to 1,500 km, then 3,000 km, and soon 6,000 km and more. This was enough to enable either side to rain nuclear bombs down on the other's territory from across the world. The term Inter-Continental Ballistic Missile (ICBM) was born.

The accuracy of the missiles was also improving. Better gyroscopes to stabilize the missiles and better knowledge of the earth's gravity resulted in targeting accuracy of better than a kilometer over inter-continental distances. Carrying a nuclear weapon that had a killing radius of several kilometers, the combination of ICBM plus hydrogen bomb meant that no one in the world was safe.

You had to keep your missiles widely separated from each other since a single strike by an enemy missile would wipe out everything within a few kilometers of the impact point. Underground ICBM missile silos were soon dotted across the great plains of America and Russia. ICBMs that could be fired from submerged sub-

---

[31] Rhodes (1995) pp. 406–407.

marines were developed by both sides. Later, the Soviets developed rail-born missile launchers, moving them around the vast interior of Russia to prevent the USA pinpointing their location.

Both sides realized that they needed to have enough ICBMs so that, even if the other side struck first, they would have enough in reserve to retaliate. And not just retaliate with one bomb, the strategy was to have enough ICBMs following a first strike by the other side to be able to destroy the latter—a situation of Mutually Assured Destruction (MAD)—the irony of the acronym was not lost on the policy makers. As both sides deployed more and more ICBMs, the number of missiles and of the bombs they carried grew rapidly. The process was called arms escalation.

By the mid-1960s both superpowers had more than enough nuclear bombs to obliterate all life on the planet several times over. The risk of accidentally triggering a nuclear exchange was recognized by both sides. Elaborate procedures were put in place to ensure that nuclear-tipped ICBMs could only be launched with the approval of the top brass—the briefcase containing the nuclear "keys" beloved of many Hollywood movies.

PARANOIA

Underpinning the arms race was a propaganda war in which each side painted the other as wanting to dominate the world. Oppressive Russian dictatorship in Eastern Europe fueled this view in America, while the USA's heavy-handed involvement in Central and South American and Middle Eastern politics reinforced the Soviet's fears. The newspaper headlines reflected the propaganda, painting all political disagreements between the super-powers in black and white terms.

In the early 1950s, a wave of anti-communist paranoia swept the USA fueled by the communist invasion of South Korea in 1950 and other events across the world, resurrecting an innate American suspicion of the extreme left that had prevailed since the 1920s, except during the brief alliance born out of necessity in World War II. The Soviets setting off their first nuclear bomb test caused a wave of panic. The Korean war was portrayed as a communist plot, which indeed it was although it is now widely accepted that the Americans contributed to widening the war by pursuing the communist forces too close to China. The arrest and trial of Soviet spies in the USA kept the anti-communist propaganda at fever pitch for months at a time.

Paranoia in the USA was given formal political blessing by the Chairman of the Congressional Committee on Un-American Activities, Senator Joseph McCarthy (Figure 13). He held a series of hearings of his committee on the subject of communist spies within the American government. McCarthy was a political pragmatist of the worst kind, using unsubstantiated (indeed, knowingly false) accusations to stir up media interest and keep his name in the news. Anyone who had ever had the slightest contact with communist organizations was black-balled from public employment and even from many private companies. Many people had sympathized with the communists in the 1930s when they were fighting Hitler—for example, in Spain—including American patriots like Ernest Hemingway. Many more had fled from

Figure 13. Senator Joseph McCarthy (1954) held Congressional hearings that fed the anti-communist paranoia in the U.S.

Hitler's Europe where they had perhaps been socialists—portrayed by McCarthy as almost as dangerous as being a communist. This hysterical movement destroyed many thousands of careers and lives.

Even after the McCarthy hearings were exposed for the sham they were, the climate of anti-communist paranoia lingered. In the 1960s the American anti-communist concern turned outward, most enduringly to South-East Asia. Communist gains in Vietnam were painted in America as the first domino in a chain of dominos. After Vietnam turned communist, next would be Cambodia, then Thailand, then Malaysia, and so on. Before long, communists would have taken over India and be at the gates of the oil riches of Saudi Arabia and Iran—so the argument went.

Paranoia inside the Soviet block was of a different form than in America. Stalin and his successors oversaw a dictatorship in which publications and broadcasts were controlled by the state. Soviet citizens knew that the official news they read or heard was at least partially false, but they had no way of knowing which parts, if any, were true. What was clear is that the state propaganda was uniformly anti-American. This "spin" on the news was convenient for the Soviet leaders since it gave them an excuse for shortages of food and other resources of daily life. They could blame the many inconveniences suffered by the masses on having to defend the homeland against America. Memories of battling Hitler could be dragged up to accentuate the dangers when necessary.

In both America and the Soviet Union the war-time atmosphere was kept alive by what became known as civil defence programs. The general public was advised on how to react to a nuclear attack. Governments built deep bunkers that were considered safe from a nuclear attack into which key public figures would descend if an attack was imminent. Many ordinary people, especially in America, also built bunkers and stocked them with food and other consumables for what they hoped would be long enough to escape the worst of the radiation.

These civil defence programs were likely to be of little use in the event of a real attack by either side. If even a small proportion of the ICBMs of either side reached their targets the whole country would probably be uninhabitable for years. A sizable exchange of ICBMs was thought likely to create the conditions of what became known as *nuclear winter*—dust thrown up by the explosions and smoke from burning forests would block out the sun for months or years, causing crop failures on a massive scale with resulting famine and disease. In some scenarios it was thought that all organic (or at least animal) life on Earth would be destroyed. A 15-megaton American test in the Pacific in May 1954 killed one Japanese fisherman and sent radioactive debris across the globe, forcing leaders in both camps to recognize the impossibility of fighting a war with such weapons—Georgii Malenkov in the Soviet Union publicly warned about the end of world civilization, while Britain's Winston Churchill noted that just a few such explosions would make the UK uninhabitable.[32]

The Soviet civil defence buildings actually made war more likely. In America, some experts interpreted the vast Soviet civil defence bunkers in Moscow and other major cities as indicating a Soviet willingness to start a nuclear war and have their citizens "tough it out" underground.

20 MINUTES TO ARMAGEDDON

The time factor was important. It would take a Soviet ICBM about 20 minutes to reach America from Siberia. So, America built a chain of radar sites across northern Canada, plus across Alaska and the Aleutian Islands to the west and in Greenland and the UK (Fylingdales in Yorkshire) to the east to give the earliest possible warning. During the most tense parts of the Cold War, a fleet of American aircraft were airborne at all times, to avoid being caught on the ground without enough time to get in the air before the incoming missiles hit. And several nuclear-powered submarines were at sea at all times able to stay submerged for weeks or months at a time thanks to their engines powered by nuclear reactors and thus immune from a first-strike attack (provided that the other side didn't know their whereabouts).

In Western Europe the time factor was really hairy. Britain and France would have less than 10 minutes (4 minutes in some scenarios) warning before incoming Soviet missiles would land. This was not enough time to reliably initiate any action, so submarine-launched missiles were seen as the best defence, since they would survive a first-strike attack.

The build-up of ICBMs and of nuclear weapons caused many inconveniences— on both sides. For one thing they were extremely expensive, and as a result many other government spending priorities had to take a back seat. Ironically, they started out as a way to save money—even President Eisenhower who saw through technical

[32] Gaddis (2005) pp. 64–65.

hype more clearly than others bought the argument that a few nuclear bombs could replace whole army regiments, pointing to the Japanese surrender after the Hiroshima and Nagasaki bombs as evidence.[33]

However, this simplistic economic thinking was soon rendered obsolete by the need to build more sophistication into military systems in response to the threat of nuclear attack. Radar networks were built to detect incoming missiles in order that the other side could launch its missiles before they were destroyed. The missiles had to be able to launch rapidly—without, for example, the painfully slow preparations we have become familiar with before a satellite is launched today. Not wanting to depend only on a relatively untried missile force, aircraft were built that had the range and the capacity to deliver nuclear bombs across the world. Submarines that could loiter undersea for months at a time then fire off missiles at a moment's notice had to be developed (Submarine-Launched Ballistic Missiles, SLBMs).

The bombs themselves were not cheap. Huge investment was needed to manu-facture the raw material of a bomb and to fabricate the precision machinery that ensured the fusion of the hydrogen at the center when the atomic bomb went off. Ultra-precise electronics is one of the critical technologies as we have learned in recent years with the publicity about illegal transfer of such devices from the US, Holland, and Germany to Pakistan and from there to North Korea, Iran, and Libya. The electronics trigger conventional explosions that drive several pieces of uranium or plutonium together to make a critical mass, thus setting off a fission bomb.

BETTER BOMBS

One way to cut costs and increase the number of bombs is to put several bombs on each missile and give each bomb a guidance system of its own so that each one hits a separate target. The multiple bombs stay inside the missile casing until its rocket motor has finished firing. Then during the 20 or so minutes (shorter if it came from a submarine) before it reaches its target, it coasts without power high above the atmosphere impelled by the enormous speed imparted by the rocket. Each of the separate bombs is ejected from the missile with carefully controlled speed and direction that will take it to its specific target. Each bomb has its own shield to protect it against the searing heat experienced as it re-enters the atmosphere in its final plunge to the target.

The ICBMs equipped with these multiple bombs were said to carry Multiple Independently-targetable Re-entry Vehicles (MIRVs). This new four-letter acronym accelerated the number of bombs deployed by each side as they replaced old missiles with new ones capable of carrying MIRVs.

Defence against ICBMs or SLBMs is difficult—impossible some said. Even if you detect an incoming missile, it is traveling too fast to hit with a conventional anti-aircraft weapon. One defence considered by both superpowers was to fire a nuclear weapon at the incoming missile sufficiently high above the target to avoid damage on

[33] Lindgren (2000) p. 30.

the ground. The early Russian anti-ballistic missile around Moscow worked on this principle.

THE CUBAN CRISIS

The fear of nuclear war prevalent throughout the 1960s reached its peak during the Cuban missile crisis in the autumn of 1962 (Figure 14). The Soviets began to install medium-range missiles in Cuba capable of carrying nuclear weapons to much of the Eastern and Central USA. The Cuban President Fidel Castro was understandably interested in having weapons that would deter an American invasion of his island—just the year before, American-sponsored forces *had* invaded Cuba and been roundly thrashed by Castro's forces in what became known as the Bay of Pigs incident, already mentioned in Chapter 2.[34]

The Soviet motivation was the same as Castro's—to provide a deterrent to American interference in Cuba, and thus underpin and strengthen Castro's communist regime.[35] Of course, it didn't hurt that having missiles in Cuba would provide some counterweight to American medium-range missiles based in various countries around the Soviet rim—in Turkey, for example. Premier Khrushchev had been complaining about these missiles for some time and he reasoned that basing missiles in Cuba would even up the threats faced by the two superpowers.[36] Once Khrushchev had decided on his course of action, Castro had little choice but to agree, since the Cuban economy was heavily underwritten by Soviet subsidies.

The CIA learned about the missile plans through their agents inside Cuba and soon American reconnaissance aircraft were filming the construction work for the missile silos and their protective anti-aircraft batteries as it progressed. The missiles themselves were then detected en route to Cuba via cargo ship.

President Kennedy received conflicting advice from his Cabinet and close associates about how to proceed—the hawks were all for sinking the cargo ships, bombing the missile sites and to hell with the consequences. Unfortunately, the consequences were thought likely to include escalation to a nuclear exchange, resulting in millions of deaths. Khrushchev too was receiving bellicose advice from some of his generals.

Kennedy finally took the advice of the doves rather than the hawks and imposed a naval blockade around Cuba, and began searching every ship bound for the island.

Around the world life seemed to slow down. Sales of newspapers and viewing figures for TV news bulletins soared as the world waited to hear the next nail-biting turn in this nuclear game of chicken. Grocery shelves were emptied as people stocked up for what might be the start of a war. Would the American naval blockade around

---

[34] Castro told French journalists the following year that his motive was not self-defence but to show solidarity with the Soviets and international socialism; Johnson (1983) p. 625.

[35] Gaddis (2005) p. 76.

[36] "Stalin would never have dared to make this move" was Khrushchev's boast according to Castro; Johnson (1983) p. 625.

Figure 14. October 18th 1962, President Kennedy (right) meets with Soviet Foreign Minister Andrei Gromyko (centre) and the Soviet Ambassador to the U.S. Anatoly Dobrynin at the height of the crisis.

Cuba stop the Russian cargo ships carrying the missiles? Would Soviet submarines sink US ships in the blockade?

The Soviet ships reached the blockade line—and stopped. The tension remained high because of the short-range missiles that had already reached Cuba and were installed there. Khrushchev negotiated what is now generally perceived as a good arrangement, whereby America agreed not to invade Cuba and to remove the American missiles from Turkey, in return for the Soviets withdrawing the missiles from Cuba.

The short-term view of the outcome was that America had won a victory— Khrushchev was ejected from the Soviet leadership two years later on the grounds that the Cuban "adventure" was a miscalculation and ended with a Soviet climb-down in the face of an American military threat. The Soviet loss of face outweighed the two significant concessions Khrushchev had extracted from Kennedy.

Recently, Kennedy has been criticized for not extracting concessions in return, such as Cuba becoming a demilitarized country similar to Finland. Cuba has indeed survived without fear of American invasion, but the anger felt by America at being forced to concede Castro's continued rule in Cuba has resulted in Cuba being treated as a pariah nation by the USA long after it has ceased to pose any kind of threat, with consequential disastrous results for the Cuban economy and the livelihood of its citizens. Two hundred years of close interaction between the USA and Cuba came to an end, and an independent Cuban government was unambiguously established.[37]

[37] Thomas (1971) p. 1418.

According to Che Guevara who was present, when Castro heard about the Soviet climb-down the Cuban leader was furious, both because of the loss of Soviet nerve and at not being consulted in the ultimate decision.[38]

The Cuban missile crisis was at the time, and is still today, considered as the most dangerous crisis mankind has ever faced. American nuclear forces were at Defcon-2, the state of readiness just below war itself, the only time this level has been declared— incidentally, the Soviets never went to full nuclear alert during the Cold War.[39] Nuclear weapons were activated on planes, missiles, and submarines. Khrushchev warned his own generals of the deaths of 500 million human beings. US Defense Secretary Robert McNamara wondered if he would live to see another week.[40]

ACCIDENTAL ARMAGEDDON

And, incredibly, two accidents almost set off WW-III at the time. You would think— well, you would hope—that at a time of crisis extra care would be taken by military forces to ensure that things did not get out of hand unnecessarily. In fact, the opposite tends to happen. Individuals become nervous, worrying that indecision and delay on their part may result in failure, and thus be tempted to press ahead in the face of unclear orders or contingencies. And, even worse, cavalier right-wing generals (or admirals) who have been advocating that their side should attack the other side anyway, might take advantage of the situation to precipitate an attack.

When the US went to Defcon 2 during the Cuban missile crisis, 3,000 nuclear weapons were put on alert (i.e., made ready for use), a total of 7,000 megatons, enough to push the earth into a nuclear winter killing most or all life on the planet, although that was not realized at the time. With the nuclear forces on a hair-trigger, a previously scheduled test launch of an Atlas missile went ahead on October 26th 1962 from Vandenberg in California landing far out into the Pacific Ocean. The test should of course have been delayed, but the Air Force commanders chose not to cancel it. If they had detected it, the Soviets would probably have taken it as a hostile launch since they would have expected Washington to alert them to peaceful launches at such a time of tension.

The Air Force Chief of Staff, General Curtis Le May, tried to bait President Kennedy into attacking Cuba. He assured the President that the Soviets would not fight back if hit hard. The president's brother, Attorney General Robert Kennedy, reported that the President had the good sense to dispute Le May's advice, saying that the Soviets would certainly respond in Berlin even if not in Cuba. It was fortunate that he ignored Le May because many years later the Soviets admitted that 20 medium-range missiles (that could reach Washington DC, for example) already in Cuba before Kennedy established the blockade were armed with nuclear warheads. Even more frightening is that a further nine short-range missiles were not

---

[38] Thomas (1971) p. 1414.
[39] Rhodes (1995) p. 576.
[40] Isaacs and Downing (1998) p. 201.

only nuclear-tipped but their local commanders in Cuba had authority to decide when to fire them—the only time such delegation occurred in the normally highly centric Soviet command structure. Hearing about this, Robert McNamara felt sure that those commanders *would* have fired the missile if attacked, following the logic of "use them or lose them".[41]

The head of the bomber and missile force, General Thomas Power, who reported to Le May, was even more gung ho to attack the Soviets, and it was Power's people who failed to cancel the Atlas launch. They went further, allowing their B-52 bomber pilots to fly beyond their holding points in the air in an attempt to provoke a Soviet attack. A U-2 spy plane even strayed into Soviet territory, causing Khrushchev to complain to Kennedy that such an incident could easily cause the Soviets to take a "fateful step". The US Navy was also over zealous and provocative, forcing Soviet submarines to surface even outside the quarantine zone that Kennedy had declared around Cuba.

Then there were the accidents. One day into the naval blockade, on October 25th 1962 an intruder was spotted climbing over a perimeter fence into a US air base near Duluth, Minnesota. Alarms at the base and at all bases throughout the region were triggered. At Volk Field in Wisconsin, the wrong alarm was triggered—the klaxon that signaled that nuclear war had begun. Pilots had been told that there would be no drills during the Cuban crisis and ran to take off in their nuclear-armed planes. Fortunately, the base commander checked with Duluth and stopped the planes as they were rolling down the runway. The Duluth intruder was a bear.

October 28th was perhaps the tensest moment of the Cuban crisis and nuclear war looked inevitable. As luck would have it, a test tape containing a simulated missile attack was being routinely run through radars in Moorestown, New Jersey, when its signals became confused with real signals from a satellite that simultaneously rose over the horizon. The operators alerted NORAD that a nuclear attack was underway and would strike Tampa, Florida, within a few minutes. Fortunately, when no such explosion took place, and other radar sites confirmed that the satellite was not a threat, the Moorestown operators realized their error.[42]

The Cuban crisis wasn't the only time when accidents happened. The previous year two 24-megaton hydrogen bombs fell from a disintegrating B-52 onto North Carolina. One parachuted down and was found. The other landed in waterlogged farmland and was never found. Examination of the recovered bomb showed that five of its six safety devices had failed.

In January 1968, a B-52 carrying four hydrogen bombs crashed into the ice a few kilometers from the American early-warning radar station at Thule, Greenland. The safety devices prevented a nuclear explosion but the conventional explosives went off scattering radioactive debris over a wide area.

An earlier crash nearly had even worse consequences. In July 1957 a US B-47 bomber crashed into a storage depot at RAF Lakenheath near Cambridge, England.

---

[41] Rhodes (1995) p. 575.
[42] Rhodes (1995) pp. 573–575 and Isaacs (1998) pp. 241–242.

Figure 15. April 1966, the missing hydrogen bomb is recovered from the Mediterranean, 80 days after a B-52 and its KC-135 refueling tanker collided in mid-air over Palomares, Spain.

The burning jet fuel was only just extinguished by fire-fighters before it would have ignited the TNT in the trigger mechanism.

Probably the most famous nuclear accident was what became known as the *Palomares Incident*. In January 1966 a B-52 strategic bomber was being refueled in the air when it collided with its KC-135 refueling tanker aircraft 9 km above Spain's southeast Mediterranean coast. Eight of the combined crew of eleven died and four hydrogen bombs fell to the ground near the village of Palomares. Although the thermonuclear devices did not go off, the high explosives in two of them did, scattering radioactive particles over several hundred hectares of farm land. A third bomb landed intact nearby and the fourth landed in the sea. After 80 days, the missing bomb was eventually found by the midget submarine *Alvin* at a depth of 750 m—part of a recovery fleet of 33 US naval vessels (Figure 15). Nearly 2,000 tons of contaminated topsoil had to be removed to the USA, and many tons of tomatoes had to be destroyed.[43]

[43] Ioanoo (1998) pp. 238–240.

A war involving nuclear weapons between America and the Soviets had been recognized as unacceptable for either side by most political leaders since the mid-1950s. The Cuban missile crisis showed that, even with that realization, events could conspire to push the adversaries almost inexorably towards a nuclear exchange. The various nuclear accidents showed that even in times of relative peace, a nuclear explosion was too close for comfort.

ALTERNATIVES TO MAD

Several political leaders of the 1960s have bemoaned their failure to find an alternative to the Mutually Assured Destruction policy—you blow up my city I will blow up yours. The idea of holding your enemy's civilian population to ransom like this was morally repugnant, but alternatives for the use of nuclear weapons proved elusive.

Robert McNamara who was Defense Secretary to both Kennedy and Johnson, told Henry Kissinger that he had tried for seven years to give the President more options (than MAD) but had given up in the face of bureaucratic opposition. Kissinger who served as Secretary of State to Presidents Nixon and Ford and before that National Security Adviser to Nixon considered that he himself *partially succeeded* in his White House years—about as close as he can bring himself to admitting failure. In 1965 "assured destruction" was defined in the Pentagon as the capacity to destroy one-fourth to one-third of the Soviet population and two-thirds of Soviet industry. By 1968 it was lowered to one-fifth to one-fourth of the Soviet population and one-half of Soviet industry. Note that in neither case did it aim at destruction of the other side's missile force.[44]

Kissinger explains the thinking thus: "the more controllable its consequences the greater the risk that a war would actually occur". Successful missile defence made the situation controllable, so made war more likely. Overwhelming superiority of one side over the other made the situation controllable, so made war more likely.

James Schlesinger, who was Defense Secretary to Nixon and Ford and thus Kissinger's Cabinet colleague, argued publicly that the US could respond flexibly to a nuclear attack, avoiding automatic escalation to assured destruction.[45] But in private he agreed with Kissinger that this stance merely increased the likelihood of nuclear war.

The other side of the coin was that the more horrible the consequences of war the less likely either side was to resort to it. Therefore, the safest policy was to aim at each other's population rather than the missiles. Kissinger was bemused by the situation where a great power found that "the invulnerability of the forces of its principal opponent, or the vulnerability of its own population, [enhanced] stability".[46]

What would happen with miscalculation was undefined.

[44] Kissinger (1979) pp. 215–217.
[45] Carter (1974).
[46] Kissinger (1982) p. 1009.

How you could defend allies in these circumstances was not clear. Allies were a two-edged sword in the MAD context. McNamara expressed concern that European allies would use their nuclear weapons to trigger the thermonuclear holocaust[47] (i.e., to drag the US into a conflict)—and he had good reason for his concern as we will see in the case of France's *entanglement* policy in Chapter 8.

The madness of MAD was that no President could make such a threat credible except by conducting diplomacy that suggested a high degree of irrationality. And who would take the moral responsibility to recommend a strategy based on mass extermination of civilians? Vulnerability of the civilian population was an *asset* in restraining the other side. For the first time a major country saw an advantage in enhancing its own vulnerability. MAD was horribly unworkable in the real world.[48]

Defense against missiles was impossible in practice—studies in the 1960s suggested that if you spent five dollars on it, the other side only had to spend one dollar to counter that.[49] However you defined the criterion, and however you varied the mix of megatons and kilotons the only way to stop the other side hitting you with nuclear missiles was to threaten the same in return—deterrence.

Use of nuclear weapons on a tactical basis was mooted. Of course, *tactical* in this context meant explosions not much smaller than Hiroshima. A 1955 NATO exercise simulated a fight with Soviet forces in central Europe in which over 300 tactical nuclear weapons were used to stop the Soviet advance. The simulation forecast a million and a half dead within two days, leading future German Chancellor Helmut Schmidt to exclaim that "tactical nuclear weapons would not defend Europe they would destroy it."[50]

Apparently, British Defence Secretary Denis Healey told President Nixon in London in 1969 that tactical nuclear weapons could be used as a warning—for example, by setting one off high in the air over the Mediterranean.[51] Healey neglects to mention this crazy idea in his memoirs, but does admit to failing to appreciate how the extremely expensive British Chevaline project was at best a waste of money. Chevaline was designed to penetrate the missile defence system around Moscow, but since the existing British missile (Polaris) could destroy the dozen next largest Soviet cities without it, UK security was enhanced not one whit.[52]

As Kissinger said, MAD was unworkable. He noted that "achieving a more discriminating nuclear strategy, preserving at least some hope of civilized life, remains one of the most difficult tasks to implement, requiring a substantial recasting of our military establishment. If unsolved, the problem will sooner or later paralyse our foreign policy."[53] It remained (and remains) unsolved.

---

[47] Healey (1989) p. 244.,
[48] Kissinger (1979) p. 216.
[49] DeGroot (2005) p. 299.
[50] DeGroot (2005) p. 191.
[51] Kissinger (1979) p. 219.
[52] Healey (1989) pp. 313, 456.
[53] Kissinger (1979) p. 217.

4

Spy satellites

In this chapter we will explore what American and Soviet spy satellites of the 1960s and 1970s could see and equally what they couldn't see. They were secret — still are for that matter—and we will see some of the peculiar effects of that. The immediate trigger for building the satellites in both countries was the "U-2 affair" which influenced the result of the 1960 American Presidential election. Having looked at that, we will see the surprisingly similar designs of satellites in both America and the Soviet Union—surprising because of the novelty and complexity of the way they got their secret images down to the intelligence analysts. After a brief look at launchers and launch sites, we will be back to the second half of the 1960s and bigger and better satellites, including the first *Big Bird*. The last of the satellites involved in the negotiations for the Strategic Arms Limitation Treaties (SALT) will then be introduced—it is also the first of what one might call the "modern spy satellites". The thorny question of whether a spy satellite can read a car number plate or a newspaper headline will then be examined before closing the chapter with a light-hearted look at how Hollywood has created a whole mythical form of spy satellites to underpin its government conspiracy stories.

But first let's discuss cell phones.

WHAT IS A SPY SATELLITE?

Satellites fly across the sky unimpeded by borders. With a camera onboard they can take photographs of the ground below (if there are no clouds in the way). Many satellites carry cameras for benign reasons such as weather forecasting or environmental monitoring, but some are explicitly seeking military information, and these we call spy satellites. If they have a suitable radio receiver they can listen in to whatever radio signals are being transmitted below, and this sort of satellite usually has a military objective. In this story the camera-carrying satellites are the main focus of the story and I will only return briefly to the radio-listening satellites at the end.

Modern spy satellites are like a camera-equipped cell phone—but with a very long telescope attached. You take a photo of an interesting scene using the built-in camera of your cell phone. The image is stored in the computer-style memory of the phone. Then you send it to whoever is interested in it via the cell phone network, in other words by radio link.

The number of pixels in the image taken by the phone-camera dictates its graininess or resolution—that is, the ability to blow it up and see further detail. The more pixels in the camera the more you can magnify it without it becoming grainy. On the other hand, the more pixels in an image, the fewer the images you can store in the memory and the more you pay to transmit it over the network. You can zoom in on the subject before taking the photo, in which case you will see more detail in the final image but less of the surroundings.

Sometimes there is no network coverage where you have taken the picture and you have to wait until your journey brings you to an area with network coverage before you can send it. If this is a frequent occurrence you may consider switching to another network that offers better coverage.

This same outline could as easily apply to a modern spy satellite, leaving aside the fact that it takes its pictures at a distance of 200 km or more through a telescope. It takes images—albeit automatically to a predetermined schedule rather than when a human touches a button. The heart of the camera is the same technology as in the digital camera or cell phone you buy in the high street—a charge-coupled device (CCD), which is a form of solid-state electronics similar to the transistor and the computer chip, which turns light into electrical messages. The images are stored in computer memory on the satellite, which can only hold a certain number of images before it reaches its capacity. The images may be radioed to ground immediately, but frequently there isn't a friendly ground station within sight of the satellite, so it waits until its orbit brings it within the coverage of its ground network.

The USA and Russia have installed extra network relay stations to improve the coverage and thus get images back to ground from more or less anywhere in the world. The relay stations are actually satellites located at suitable, very high orbits which relay the images from the spy satellite to the relevant ground network. These satellites therefore have no intrinsic limitation as to the number of images they take, provided the radio link has enough bandwidth to carry them. Other spy satellite operators such as France, China, and India have to wait until their satellite appears over their ground stations before receiving the recorded images.

However, the digital camera is a recent innovation, and the spy satellites that follow the same principle are also relatively new. Until recently a camera involved a roll of film which was *exposed* one scene at a time. The resulting exposed film roll then had to be *developed* at a specialist shop, which involved chemically processing the film in the absence of light. This turned the film into a *negative*, and prints were then made of each scene on the negative to provide the traditional photographs. The digital camera has eliminated the roll of film and its processing. Now prints are made from a digital storage device such as a CD or a memory card, from an email attachment, or from a shared web store.

The first spy satellites, too, involved a roll of film which had to be processed chemically under carefully controlled conditions. Both the Americans and the Soviets initially hoped to process the film onboard the satellite, scan the resulting negative and radio the result to ground. The idea was backed in America by Edwin Land, the inventor of the Polaroid camera that processes its own films automatically and instantly. President Eisenhower, who was a self-trained photo-interpreter,[54] had appointed him to advise on a number of reconnaissance technology matters, and the idea of a spy satellite incorporating the camera scan/radio concept was approved in 1955—two years before the launch of the first Sputnik. Eisenhower was the first US President to understand how useful aerial intelligence and satellite photography could be. He had gained this experience using photos taken high over Nazi Germany by reconnaissance aircraft to plan the movement of the forces he commanded as Supreme Allied Commander during World War II.[55]

Land was a Harvard drop-out who invented the instant photograph in the 1940s in response to the question from his daughter Jennifer "Daddy, why can't I see the picture now?" as he took family snapshots at Christmas 1943. It took him three years to work out how to do this, until in February 1947 he presented the instant camera and its Polaroid film at a meeting of the Optical Society of America. A year and a half later the first black and white model went on sale—the color model had to wait until 1964.

Land wanted reconnaissance satellites to "see it all, see it well and see it now." See it all meant wide area coverage, see it well meant high resolution, and see it now meant a system that got the images to the photo-interpreter in minutes not days. Processing a film onboard a satellite was possible and was in fact achieved by both America and the Soviets in the early 1960s. But the radio link to ground had insufficient bandwidth to get enough full-quality images back. You can compare this problem with that of narrowband versus broadband access to the internet. Downloading a high-quality image takes what seems like forever via a narrowband link, but becomes routine and fast when broadband is available. The 8 MHz available on the early American satellites was the equivalent of narrowband in this respect, and the whole idea of developing and scanning the films inside the satellite then transmitting them via radio link to earth was abandoned for about 20 years by both super-powers.

The film scan/radio downlink technology in fact first appeared in the deep-space missions of both countries, being used first by the Russians for their Lunik moon probes and by America for the Mariner missions to Venus, Mercury, and Mars, and later for the highly successful Lunar Orbiter series of satellites that mapped the moon in preparation for the Apollo landings. The first successful spy satellites incorporating image delivery via radio links to ground had to await further developments in electronics, in particular the now ubiquitous CCD mentioned at the start of this chapter.

[54] Temple (2005) p. 33.
[55] Isaacs and Downing (1998) p. 353.

So, in both countries a less ambitious concept was actually used for the spy satellites of the 1960s. The scheme eventually adopted by both America and the Soviet Union was for the exposed but unprocessed film to be ejected from the satellite in a capsule that returned to earth—in the American case caught by an aircraft as it descended on its parachute over the ocean, while the Soviets brought theirs down directly on land. Once back on earth the film was rushed to a film-processing laboratory to be turned into useful pictures.

When the film is taken from a normal camera and sent off for processing, the user inserts a new film into the camera, ready to take more pictures. Reloading the camera inside the satellite wasn't a practical proposition at first, because each film would need a separate return capsule. The return capsules were heavy and bulky and the first spy satellites could only carry one—later American satellites carried two or three capsules and the film could be cut at any point to be put into the first capsule while the rest of the film remained in the camera to continue being exposed and eventually returned in the final capsule.

The Russian Zenit satellites had a variation on this, in that they returned the whole camera not just the film. This meant that the capsule was bigger and bulkier, but it avoided the need for an automated mechanism to extract the film from the camera onboard the satellite, and it allowed the camera to be re-loaded with film and re-launched on a future satellite, thus saving money. In the 1970s the Soviets introduced the Yantar series of satellites that had three return capsules, in the last of which the camera itself returned as for Zenit.

These days, film-processing shops will return your prints within one hour—or even less. But such rapid turnaround is a relatively recent proposition—until the 1990s it was common to wait several days before the prints and the associated negative came back. The same was true with spy satellites. In originally favoring the camera scan/radio concept, Edwin Land had wanted to avoid the delays inherent in the capsule approach—but it would be the 1970s before a radio link replaced the capsule.

Frequently, the capsule came down thousands of kilometers from the processing laboratories, probably far out to sea. Getting the film back to civilization could take days, and then it had to take its place in the queue at the processing lab. Having been processed the film negative had to be copied and two or more positive prints made before the photo-interpreters got their first look at the results. And about half the time the result was nothing but a picture of clouds with no ground visible!

Both the Americans and the Russians had many failed attempts before they got their respective spy satellites to work. The Americans found, for example, that the standard film they used didn't work in the vacuum of space—the acetate-based film developed by Eastman Kodak for the U-2 spy plane which worked fine at 25 km altitude became brittle 135 km higher in the hard vacuum of space. The film also had to negotiate various twists and turns in its passage through the camera due to the small size of the satellite (Figure 16). Vacuum causes many materials to "degass", as it's called, whereby gases and liquids that remain stably within a substance on earth boil off when it is exposed to a vacuum. The oils used to lubricate the mechanical parts of the camera and the glues and waxes used to seal it were obvious sources of

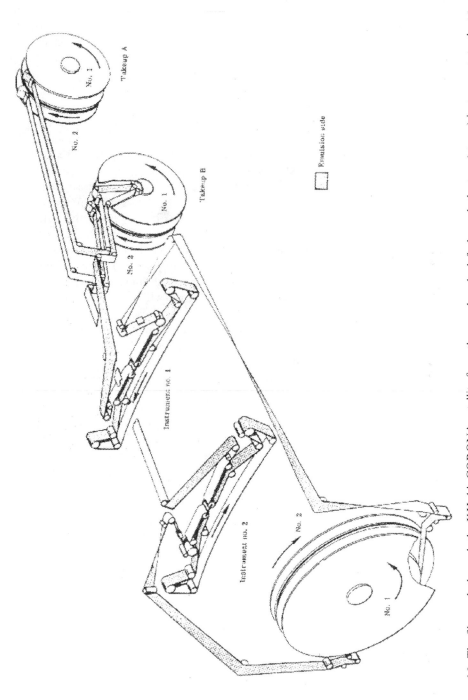

Figure 16. The film path through the KH-4A CORONA satellite from the spool at the left through the two (stereo) instruments to the two take-up reels on the right—one for each of two return capsules. Credit: Dwayne A. Day.

degassing, but in fact most substances experience the effect to some extent. Combating this phenomenon is now a well-understood activity in space projects, but in the late 1950s very little was known about the sensitivity of different materials to the condition. After some clever design and development, Eastman Kodak devised a new film on a polyester backing, apparently requiring some breakthroughs in the chemistry to make the high-resolution emulsion stick to the polyester.[56]

The Russians avoided most of these degassing problems by sealing the camera from the vacuum and maintaining the air pressure inside. This made for a bulkier and heavier satellite but greatly simplified the development of the cameras and telescopes it carried. The less complex parts of these Zenit satellites were open to the vacuum such as batteries, radio equipment, orientation and braking rockets, and environmental control (heating and cooling). Weight and bulk were not as severe a constraint for the Soviets because of their much larger rockets. The American CORONA satellites weighed about a ton, rising by the mid-1960s to about one and three-quarter tons as the rockets gradually improved. The early Soviet Zenit satellites tipped the scales at almost 5 tons and were 5 meters long. The first Zenit contained two cameras and this soon increased to four—one took a wide-area image to show the orientation and location of the satellite, while the other camera(s) took high-resolution pictures. Unlike the American panoramic (almost fish eye) images which were highly distorted near the edges, Zenit took photos directly onto flat frames resulting in relatively undistorted images. This made such advances as stereo imaging relatively straightforward to introduce.

The Zenit carried a sufficiently large number of frames to take images of the whole of the USA—similar to the capability of the American CORONA that used the more compact film rolls.

Some people had doubted that catching the parachute-borne capsule in mid-air as it descended at about 25 km/h would ever be possible, but in fact this ingenious feat was achieved with relatively few hiccups (Figure 17). Night-time recoveries of illuminated capsules were even possible.[57] A few fell into the water where they floated for about a day before a salt plug in the base dissolved and they sank to avoid falling into the wrong hands—broadcasting a homing radio signal and flashing a light while afloat to help the recovery teams find them.

The Soviets avoided this whole problem by bringing their capsules down over land, and allowing the return capsule to soft-land without being caught in mid-air.

SECRECY

The American spy satellite program was initially relatively well-publicized. However, once Sputnik had been launched and the intense public interest in the subject became apparent, President Eisenhower realized that he had to remove it from the public eye. The result was to have unexpected, unfortunate, and somewhat bizarre consequences

[56] Day *et al.* (1998) p. 56.
[57] Day *et al.* (1998) p. 60.

Figure 17. A C-119 snares a parachute-borne CORONA film return capsule in mid-air over the Pacific Ocean Credit: National Reconnaissance Office

for two of the system designers. A public announcement was made that the spy satellite project was being canceled, including cancelation of the contract with Lockheed to build it. The two men from the RAND Corporation who had originated the whole idea, Amrom Katz and Merton Davies, were furious. Katz in particular lobbied hard and publicly for the project to be reinstated. The project in fact *had* been reinstated under another name, but once Katz went public with his complaints the military had to keep up the pretence. Katz therefore lost his security clearances to work on such ultra-secret or *black* projects, and his and Davies' careers were damaged. After a year or so they noticed that their former colleagues wouldn't speak to them about spy satellites and realized that the project was going ahead secretly and without them—a bitter pill indeed to swallow.[58]

The contributions of Katz and Davies were eventually recognized by the government in September 2000 when they were among the 10 people awarded the title of Founder of National Reconnaissance. The citation accompanying this accolade mentions their roles as advisors to President Eisenhower, helping to make the strategic reconnaissance policy a success, providing technical expertise to shape the emerging discipline of national reconnaissance, and underpinning the founding of the National Reconnaissance Office (NRO) in 1960–1961.[59]

Incidentally, the 10 Founders and 46 Pioneers that made up that first intake in 2000 to the NRO hall of fame were all men. Perhaps the secrecy of the programs mitigated against female participation, because in the civilian programs like Apollo women were certainly present. The NASA publicists made much of my TRW colleague Poppy Northcutt who worked in mission control across the street from my office—"a beautiful star to guide them home" was one of the phrases the press picked up on. Poppy herself was the first to point out that the job was very much a team effort, but men in white shirts and ties tended to get pushed into the background when Poppy's stunning blonde good looks were in evidence. The BBC's then aerospace correspondent, Reg Turnill, often reminds me of the times he met Poppy, and she gets

---

[58] Day *et al.* (1998) p. 113.
[59] McDonald (2002) pp. 359–360.

several mentions in his recently published memoirs.[60] Perhaps the software end of the business where I work is atypical, but Poppy was by no means the only glamorous female making important contributions to Apollo. For example, in the next office to me at TRW was Cissy Phillips who introduced me to the challenges of determining the orbit of a satellite around the moon, and down the hall was Trilla Di Sulima, a key member of the group that developed the software my team used to analyse Apollo navigation problems.

Having come to grips with Voltaire's injunction that "the best is the enemy of the good" and deciding to settle for a good surveillance system rather than the perfect one, the revised and less ambitious project was given the name CORONA, and at Eisenhower's explicit command was given to the CIA to run instead of to the Air Force. The CIA had a modest amount of experience at managing expensive hi-tech projects but Eisenhower simply didn't trust the military to keep these programs secret; his experience told him that eventually they would be tempted to leak information to Congress when it came time to argue for an increased budget appropriation. No doubt he had benefited from this tactic during his long military career, and he had certainly witnessed examples since becoming President. The name CORONA is said to have been whimsically assigned based on either the well-known brand of cigar or typewriter that the CIA's Richard Bissell's eye happened to rest on when the question arose.[61] A cover story was created in which the Air Force began a Discoverer program to place biological specimens in orbit and recover them for analysis. This largely fictitious program was said to be aimed at helping understand the difficulties of manned spaceflight. It gave a convenient cover for CORONA launches and recovery activities.

The Soviet satellites were always kept secret. To make life as difficult as possible for a satellite's purpose to be worked out by foreign analysts, all Soviet launches were given the Kosmos designation, each distinguished from the previous launch by a sequential number. The *Zenit* program (including its name) was a closely guarded secret, but curiously Zenit was also the name of a popular brand of SLR camera in the Soviet Union made by the same factory that made the spy satellite camera.

THE CATALYST—U-2 AND THE MISSILE GAP

May 1st 1960 was a watershed for the American policy on spy satellites. Before that date America had been flying U-2 planes over Russia to photograph its military facilities. The Russians were perfectly well aware of these over-flights but were powerless to stop them as their anti-aircraft missiles did not have the required range. Until, that is, May 1st 1960.

That morning Gary Powers, a veteran of 27 previous flights, took off from the U-2 base in Turkey intending to fly across Russia to Norway taking photographs of two ICBM test sites, at Sverdlovsk and Plesetsk. The Soviets fired a volley of SA-2

---

[60] Turnill (2003) p. 228.
[61] Day *et al.* (1998) p. 273.

anti-aircraft missiles at the U-2 and somehow brought it down. Powers ejected and descended on his parachute spending the next two years in a Soviet jail. There is some doubt about how exactly the U-2 was destroyed, since the SA-2 should not have been able to reach its 20 km altitude. One view is that the shock waves created by the exploding missiles were enough to damage the fragile U-2.

The downing of the U-2 came just as President Dwight Eisenhower and Premier Nikita Khrushchev were to meet in Paris for their first Summit talks. The meeting broke up in acrimony when Eisenhower refused to apologize for the incident.

Thereafter, over-flights of the Soviet Union by the U-2 were out of the question, leading to a gap in American visibility of Soviet ICBM deployments until the first spy satellites were successfully deployed. The two main candidates in that year's American Presidential election argued vigorously over the "missile gap" that was thought to exist between the allegedly extensive Soviet ICBM fleet and that of the USA. Democratic candidate John F. Kennedy made much of the lack of information following the shooting down of the U-2, which his Republican opponent, Richard M. Nixon, was unable to counter effectively. The apparent Soviet lead in spaceflight also played to Kennedy's advantage, although it too was largely a myth (see Chapter 1).

THE EARLY SATELLITES

To stay in orbit a satellite has to be traveling at about 29,000 km/h, otherwise it falls to earth. They also have to be at 160 km (100 miles) altitude or more, otherwise the drag of the atmosphere slows them down too quickly. Indeed, even if traveling with the required speed, the tenuous atmosphere at 160 km altitude still slows them down slightly but inexorably, so that after a few weeks or months their speed falls below that needed to stay in orbit and they descend into the lower atmosphere burning up due to their high speed.

At 29,000 km/h, a satellite is crossing the ground at 8 km/s or 10 to 30 times faster than a rifle bullet. The cameras in the satellites took pictures of the strip of land below the satellite as it traveled around the globe. The first American camera was a variation of the wide-angle camera used to take panoramic or long rectangular pictures of, for example, a school class. Each picture was an image 16 km wide in the direction the satellite was moving, and 190 km wide from side to side. Then the film was moved forward just like you do in a conventional film camera ready for the next exposure— 2 seconds later and 16 km ahead of the previous image. This series of 16 km-at-a-time images built up into a strip many hundreds or thousands of kilometers long by 190 km wide. For the brief moment that the exposure was being taken the blur of the satellite's movement over the ground was compensated by mechanically moving the camera lens in the opposite direction. The satellite was programmed when to start and end each strip, timed to coincide with its passage over the area of the earth's surface of interest. And about half of the time the ground was obscured by cloud cover in the areas of most interest.

The CORONA satellites soon included other small cameras ("sensors" as engineers like to call them) to show the horizon and the stars to enable analysts to

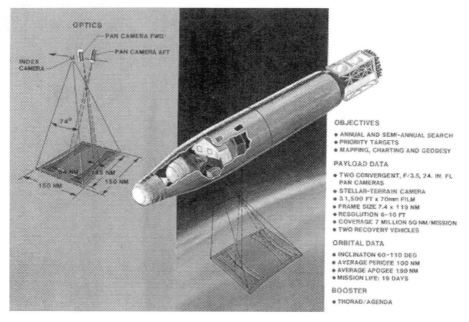

Figure 18. The US CORONA satellite was attached to the Agena upper stage of the Thor-Agena launcher; the Agena provided attitude control, battery power, and thermal protection for CORONA. Credit: National Reconnaissance Office.

work out the direction that the satellite was pointing at all times and thus determine the location of objects in the images.

The Soviet Zenits had a slightly different camera design involving square frames rather than a film roll, but otherwise the concept was similar. The Zenits could tilt the satellite to point the camera at a region of interest, a facility not initially available on CORONA.

The satellites had to be carefully stabilized to ensure they maintained a steady pointing direction and avoided vibration that would blur the images. The required stability proved difficult to achieve, and both the Soviets and the Americans took advantage of developments made for other programs. The Americans used the left-over upper section of the rocket on which the satellite was launched (Figures 18 and 19). The very first rocket pioneers had recognized that to get the maximum range or speed you needed to drop off empty fuel tanks as you rose higher and higher—avoiding the unnecessary weight of the now useless tanks. Rockets would therefore have two, three, or even four *stages* with only the final stage entering orbit—the others falling back to earth and either hitting the ground or burning up in the atmosphere due to their high speed—launches tended to be from coastal sites and directed out over the ocean or in sparsely populated areas of the world in order to avoid the risk of empty stages landing on populated areas. The rockets had to be carefully stabilized to stay on course, so the Americans decided to build the camera system and the return capsule into the last or Agena stage of the Thor-Agena rocket.

Figure 19. One of the early CORONA satellites undergoing pre-launch vibration tests at Lockheed's facility. Credit: Dwayne A. Day.

The Soviets built a stand-alone satellite that was placed in orbit by their last rocket stage, but used the same basic satellite as they developed for carrying first animals and then humans into orbit, the Vostok satellite. The spy satellite, called Zenit, turned out to require a much more sophisticated stabilization system than the manned Vostok capsule and this took some time to design. The stabilization design in fact owed more to that developed for the headline-grabbing Luna-3 satellite that took the first pictures of the moon's farside[62] in 1959.

The early American and Soviet satellites showed objects as small as about 6 m in size, but the cameras were gradually enhanced so that by the mid-1960s this improved to about 1.5 m.

The first successful launch of an American CORONA spy satellite took place in August 1960, returning pictures of huge swaths of the Soviet Union. The resolution of about 10 meters showed air fields and other major installations but did not allow individual weapons or radars to be characterized. The wide photographic coverage achieved with this first satellite was enormous in comparison with the single strip covered by a U-2 flight. The photos showed Soviet military facilities in eastern Siberia clearly. However, photos suitable for detailed military analysis had to await the launch of cameras with better resolution than this first flight. We will come back to what these satellites could detect in the next chapter.

Both the satellite and a U-2 aircraft took images showing a swath of ground about 190 km wide, but the satellite orbited the earth every 90 minutes, thus making it possible to cover the whole Soviet Union within a day or two (Figure 20). The U-2 flights took just a single strip of images and then might not take another for several months. The first satellite took images of more of the Soviet Union than all previous U-2 flights combined!

The importance America attached to getting these satellites into orbit can be gauged from the fact that this was the first success after 12 failed attempts, starting in March 1959, 15 months before the first successful flight—either the launchers had blown up or the satellites or recovery capsules had failed. The Soviets too had a dozen failed efforts before their test program was complete. They successfully recovered a pair of dogs in August 1960 which augured well for film capsule return, but it wasn't until August 1962 that the first Soviet pictures were successfully returned—two years after America—and October 1963 before their test program was wound up and Zenit became operational.

The first Soviet satellites, the Zenit series, provided pictures with a resolution reported to be about 10–15 meters—similar to the first American CORONA satellites. Confusingly, the resolution of the pictures was said to be good enough to count the number of cars in a car park, which suggests a resolution more like 3–5 meters.

Both sides rapidly improved the performance of the cameras onboard the satellites. The Soviet Zenit-3 first launched in 1963 had a resolution of 1–2 metres. The American CORONA satellites were also steadily upgraded and by the mid-1960s

---

[62] The moon keeps the same face towards us at all times, leaving the other side forever unseen from earth.

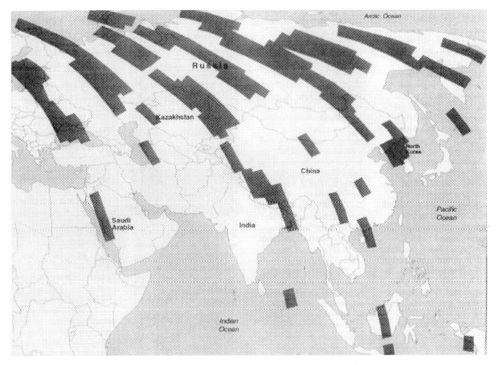

Figure 20. Areas photographed by a CORONA satellite on a typical mission. Credit: National Reconnaissance Office.

had a camera with 1–2 meter resolution. In addition, the USA GAMBIT satellite was launched from 1963 onwards to take close-up photographs of specific targets with a resolution of better than 50 cm.

A satellite flying over Russia or America takes a series of photos in a strip which is typically a few hundred kilometers wide. When designing the satellite, the designers have to decide whether to increase the width of the swath but at the expense of the resolution of the details within it. Likewise, the satellite could be designed to enhance the resolution of the photographs allowing finer details to be seen but at the expense of narrowing the width of the swath. This latter type of surveillance is often likened to observing a scene through a drinking straw—you can see great detail of a very narrow area.

The Soviets from very early on flew a special camera in each satellite to take wide-area pictures while the main cameras took the high-resolution images. The wide-area images helped to clarify where exactly the main camera was pointing. From 1963 the Americans flew the two different types of camera in two separate satellites—CORONA for the wide area and GAMBIT for the close-ups—it wasn't until the early 1970s that they had this twin capability in a single satellite.

Another issue for both sides was the need for clear skies—and for daylight. The cameras used were unable to see through clouds, and large parts of the Soviet

Union are covered in cloud for long periods of the year. Likewise, at least initially the cameras needed daylight to light up the scene to be photographed. Later satellites carried cameras sensitive to infra-red radiation, which allowed them to capture images of heat-emitting bodies. However, the resolution of infra-red images was never as good as that of optical imagery, and they also lack the contrast of visible light images, thus making it difficult to characterize objects.

Besides providing night-time pictures, infra-red imagery can also spot features not evident in normal light. For example, underground constructions will have a different heat signature from the surrounding area, thus showing up in the infra-red. Just as visible light comes in many different colors depending on the frequency of the light, so too infra-red comes in a range of frequencies which can be thought of as infra-red colors. Many chemicals show up strongly at specific infra-red frequencies, so by taking photos at several infra-red frequencies a photo-interpreter may be able to identify the chemical composition of an object.

An important enhancement introduced in the American CORONA satellites from February 1962 was the ability to image in stereo. A satellite photo in a populated area may contain considerable detail that is easy to recognize—roads, bridges over rivers, fields, etc. But in barren countryside, it is difficult to make out any specific features from the pictures. Small buildings are often hard to distinguish from general changes in soil discoloration, for example.

Stereo imagery changes that dramatically, making buildings stand out sharply from the surrounding terrain, and ditches and fences emerge from what look like squiggles on the mono image. Stereo imagery had been available in the U-2 images thanks to a clever twin-film camera system that simultaneously exposed two slightly different views of the ground, and was warmly welcomed by the photo-analysts when satellites finally provided it. Satellites are severely constrained as to the weight they can carry, and also as to the volume or bulkiness of the equipment inside. Electrical power is also often a limiting factor, so in addition to building a camera that works in vacuum the engineers had to miniaturize everything to make it compact, and use light-weight materials. Then it had to work unattended for several days or weeks without fail. So, designs that fitted inside a U-2 had to be radically changed to work in space. The CORONA stereo MURAL or "M" camera had two film strips supplied by a cassette which diverged to enter the two cameras, then bent back to enter the return capsule and be rolled on to twin cassettes. The two cameras pointed at slightly different angles so that they photographed the same scene half a dozen or so frames apart, and thus from a different direction. When photos of the same scene from both cameras were properly aligned and viewed on the ground through a stereo-microscope they appeared three-dimensional.

Another important feature introduced in the mid-1960s was the ability to point the camera at a target of interest. You can make a satellite over-fly a target of interest either by changing the orbit of the satellite to pass over the target or by pointing the camera onboard the satellite off to the side towards the target. It seems obvious that pointing the camera would be preferred, but experience shows that a mechanical device is one of the things most likely to fail on a satellite. Hence satellite designers hesitated before putting a mechanism inside the camera that would allow it to look to the side. Eventually such cameras *were* built and provided a useful degree of flexibility to the analysts interested in a specific target. The disadvantage to taking a photograph off to the side is that the resulting image will be taken at an oblique angle and may therefore be difficult to interpret—not only will the shapes of objects be distorted but the distortions introduced by the atmosphere will be greater. As mentioned previously, the Soviets had this feature in their Zenit satellites from the beginning.

The orbit of the early satellites tended to be slightly elliptical having a minimum altitude of about 200 km and maximum of about 400 km. A more nearly circular orbit would have simplified the camera design. When the camera shutter was opened briefly to take a picture, the ground below moved slightly even in the very short period of the exposure. Image motion compensation was designed into the cameras to counteract this, which otherwise led to blurring of the image and loss of fine detail. The amount of motion compensation required varied depending on the speed of the satellite, which in turn depended on the altitude above the ground. A circular orbit would always be at the same altitude and so the motion compensation could be simpler. One of the factors in the steady improvement of image quality of the later CORONA and Zenit missions was the advances in the image motion compensation techniques to accommodate an orbit with varying altitude. You may be familiar with effectively the same technology in your digital camera or camcorder under a name that varies from brand to brand—for example, anti-shake, vibration reduction, steady shot, or image stabilization.

The first CORONA had a very simple motion compensation scheme that assumed the orbit was circular, explaining at least in part its relatively poor resolution. The next few missions had an improved camera that could vary the motion compensation to allow for the effects of the elliptical orbit, resulting in image resolution of about 6–8 m. The September 1961 mission was the first to fly a much different camera incorporating not only variable motion compensation, but also other vibration-reducing design changes. The result was images with resolution described as 3–7 m.[63]

We are used today to computers controlling actions and events in almost limitless flexibility. I think, for example, of the software my colleagues at Logica developed to control the descent of the Huygens probe through the atmosphere of Titan (Saturn's largest moon) down to a soft-landing on the surface in January 2005. Earth was more than a billion kilometers away, so that a radio signal from earth took nearly an hour to reach the probe. Control from earth was therefore totally impractical since the whole descent lasted only 2 hours. The software in the probe had to respond to information about itself and Titan (pressure, wind speeds, acceleration, etc.) and execute various critical actions accordingly.

I mention this to compare with the relative simplicity of the control logic on the first CORONA satellites. The earliest ones had a pre-programmed sequence of holes punched into tape that unwound at a fixed rate. The tape passed over five metal contacts, and depending on which contacts and how many of them were activated at

---

[63] Day *et al.* (1998) p. 65.

any given time, various events would be triggered on the satellite, such as start or stop the camera, or eject the recovery capsule. Initially, the only control the ground had was to advance or retard the timer. There was no way to change the sequence of events which made for a very inflexible setup—typically the satellite orbit did not follow exactly the planned orbit, and after 2 or 3 days would be far from the area assumed when the tape was created.

After about a year, the satellite was improved to allow commands radioed from the ground that could control the camera directly. Targets for the camera could be changed, and information about cloud cover could be used to switch the camera on or off.[64]

The early satellites stayed in orbit for only 3 or 4 days, by which time they had used up their complement of film. For such a short duration, the orbit could probably have been even lower than the typical 200×400 km used without reaching the critical point where air drag becomes strong enough to pull it down and burn up in the atmosphere. As the satellites got more sophisticated and longer mission times became possible—in the case of CORONA, for example, by virtue of the introduction of the second and third return capsules (Figure 21)—the orbit had to either become higher or the satellites had to carry fuel to adjust their orbit from time to time and stop it falling below the critical point.

The elliptical orbit meant that the images varied in quality depending on where the satellite was in the orbit—at the point of closest approach to the earth, the *perigee*, the images were best, while at the farthest point from the earth, the *apogee*, they were considerably less detailed. In a 200×400 km elliptical orbit the difference in detail was a factor of 2. This feature of the orbit was turned to advantage by both America and the Soviets. The low part of the orbit was invariably in the northern hemisphere where all the important targets lay, while the higher end of the orbit was in the much less interesting southern hemisphere. So, the orbit lifetime was longer than for a 200 km circular orbit with no loss of useful images.

LAUNCHERS AND LAUNCH SITES

The number and frequency of launches seems stunning by comparison with today's satellites. Thirty-eight of the GAMBIT KH-7 close-look satellites discussed below were launched in the four years they were used (July 1963 to June 1967) or about one every 5 weeks. The CORONA wide-area survey satellites were even more prolific, launching a new satellite every 4 weeks during the first six years, even excluding the first 12 failures and the various mapping, geodetic, and other non-standard flights.

The CORONA Technical Director at the CIA, A. Roy Burks, recalls that 12 complete CORONA payloads had to be in storage at all times, and launch of one had to be possible within 30 days.[65] By comparison, one or two spy satellites a year is about the going rate today. Of course, the technology has radically changed—the

---

[64] McDonald (2002) pp. 277–278.
[65] McDonald (2002) p. 175.

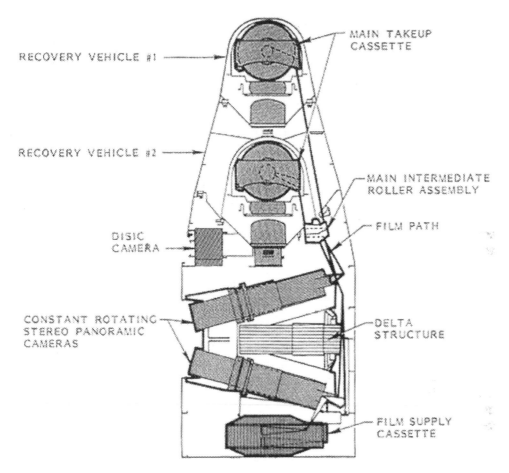

RECOVERY VEHICLE #1

MAIN TAKEUP
CASSETTE

RECOVERY VEHICLE #2

MAIN INTERMEDIATE
ROLLER ASSEMBLY

FILM PATH

DISIC
CAMERA

CONSTANT ROTATING
STEREO PANORAMIC
CAMERAS

DELTA
STRUCTURE

FILM SUPPLY
CASSETTE

Figure 21. Schematic of CORONA showing the film cassette at the base, then the twin stereo cameras, and finally the two recovery capsules at the top. Credit: National Reconnaissance Office.

early satellites were only useful while their film lasted, whereas today's satellites use digital imaging which never runs out of film!

Just like in the Soviet Union, the US launchers all had a military heritage. The Thor was originally developed as an Intermediate Range Ballistic Missile (IRBM), the Atlas was the first US ICBM, and the Titan was another ICBM. The Atlas launchers were in fact decommissioned ICBMs which had been replaced in that role by the more resilient Titans.

The launch sites, too, were often a mix of civil and military—although in the Soviet Union pretty much everything was military. The world's busiest launch site has historically been Plesetsk in Russia—a large proportion of its launches being spy satellites of one form or another, or missiles. Its northerly location (latitude 63°)

allows it to launch satellites towards the north or northeast which is required for the near-polar orbits of most spy satellites.

The US equivalent of Plesetsk is Vandenberg Air Force Base in California from where launches into polar orbit go due south out over the Santa Barbara Channel. Situated 200 km northwest of Los Angeles, highway US # 1 takes the motorist inland of the base through the Lompoc Valley. But the railroad goes right through the base, and the launch countdown has to be put on hold as trains go through. During his 1962 visit to the USA, Soviet Premier Khrushchev was a passenger on a train that passed through Vandenberg, but there is no record of his taking advantage of it for intelligence purposes.[66]

The other major Soviet launch complex is Baikonur as it is generally known in the West or Tyuratam as Kazakhstan officially calls it. The word "complex" is used to emphasize the size of these Soviet bases that stretch over a hundred or more kilometers from one side to the other.

The Soviets refused to publish the location of their launch sites for many years, and they were off limits to Soviet citizens and of course to foreign visitors. However, Newton's laws of motion make it fairly easy to work out the approximate location of where a rocket has been launched, and the location of Plesetsk was worked out by an English high school teacher, Geoffrey Perry, immediately after the first launch from it in 1966. The Soviets then engaged in misinformation to protect the location of their other main launch site, Baikonur, giving as its location the village of that name 400 km to the northeast of the actual base. The launch site in fact is 200 km east of the Aral Sea in southern Kazakhstan close to the Moscow–Tashkent railway. Its true location was worked out for the general Western public (the intelligence services of course knew exactly where it was) by a Japanese astronomer in 1957. Baikonur is where all of the Soviet, and now Russian, manned launches occur, but most of its launches are military—satellites and missiles. In recent years as Russia has joined the capitalist world, it is increasingly used for commercial launches.

The best known launch site in the world is of course the US complex on the east coast of Florida near the small town of Cocoa Beach and the beaches and lagoons of Cape Canaveral. The Eastern Test Range as it was known in the 1950s was associated with Patricks Air Force Base, and sent missiles eastward and southeastward into the Atlantic Ocean. The NASA facility at Cape Kennedy is what most people think of as the home of America's space program, and in the early days of the space age it was home to many of the most visible American attempts to reach orbit. However, from Cape Canaveral it is unsafe to launch directly northward (veer off course and you hit Washington DC or New York) or southward (Caribbean islands and South America), so the polar orbits most suitable for spy satellites have to be launched from Vandenberg in California. So, while the public's eye was on Cape Canaveral watching early US rockets explode and early astronauts turn into heroes, the real space race was happening on the West Coast.

[66] McDonald (2002) p. 314.

SATELLITES GET BETTER—AND BIGGER

Figure 22 illustrates the growth in size and complexity of the US CORONA satellite throughout the 1960s. One of the most important enhancements was the improved camera introduced by the USA in 1967 on the so-called KH-4B variant of CORONA—designated as J-3 in Figure 22. It had improved motion compensation that allowed the lowest point of the orbit to drop to 130 km. It also had less jitter and the resulting images under the best conditions had a resolution of about 1.8 m—and a stereo capability.[67] Because these satellites had two recovery capsules, they would spend two to three weeks in orbit effectively parked and waiting to be maneuvered to a specific target. Their orbit would be modified to bring the perigee directly over the target area and the perigee would be lowered to between 120 and 130 km. The orbit would then be stabilized to stay over that target for a few days—"stay" in the sense that during a few of its 14 orbits each day, the satellite came over the target for a few minutes. Having completed its visit to that target the orbit would be raised to increase the orbital lifetime and await the designation of the second target.

The CORONA KH-4B was designed to take wide-area pictures and thus ensure complete coverage of the Soviet Union and other countries of interest as often as possible. The US also had a family of satellites that took more detailed pictures of a narrower area, the GAMBIT series. The strip of ground that CORONA photographed was 290 km wide (early versions were 190 km) while that of the initial GAMBIT KH-7 family was only 22 km wide. GAMBIT was intended to take close-up pictures of targets—many of which would have been identified in CORONA images. The resolution of GAMBIT KH-7 images was about 50 cm (about 1 m in the first few flights), significantly better than the 1.8–5 m of the best of the CORONA series, the KH-4B.

Although superficially similar in shape to the CORONA satellites the GAMBIT KH7 was considerably heavier, weighing at least 2 tons (not much information on GAMBIT has been declassified yet, and that figure comes from the publicly available TRW Space Log, so is probably on the low side). It required the Atlas rocket to get it into orbit, quite a bit more expensive than the Thor used for the CORONA KH-4B.

The US satellites kept getting bigger and bigger. The GAMBIT series had a major upgrade in 1966 with the launch of the first of the 3-ton KH-8s. These required the even more expensive Titan 3B launcher to get them into orbit, but still averaged an impressive launch every 6 to 10 weeks through 1970.

The bigger and better GAMBIT KH-8 returned images with a resolution of about 8–10 cm and initially stayed in orbit for about 10–12 days which was a few days longer than its predecessors. It had at least two and perhaps four return capsules to give flexibility in sending back timely information.[68] From 1969 the mission duration rose to 20 days and continued getting longer until the final missions in 1973–1984 lasted 30, 40, and eventually over 100 days. The longer mission times were no doubt partly due to carrying more film and extra return capsules, and having more

---

[67] Day *et al.* (1998) pp. 80–82.
[68] Burrows (1986) pp. 234–235, and Lindgren (2000) p. 111.

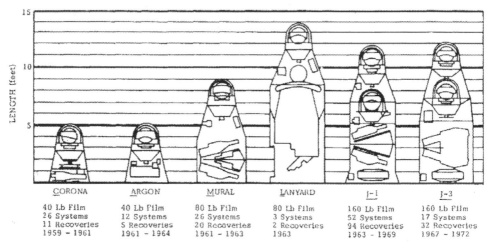

CORONA	ARGON	MURAL	LANYARD	J-1	J-3
40 Lb Film	40 Lb Film	80 Lb Film	80 Lb Film	160 Lb Film	160 Lb Film
26 Systems	12 Systems	26 Systems	3 Systems	52 Systems	17 Systems
11 Recoveries	5 Recoveries	20 Recoveries	2 Recoveries	94 Recoveries	32 Recoveries
1959 – 1961	1961 – 1964	1961 – 1963	1963	1963 – 1969	1967 – 1972

Figure 22. The growth of the CORONA satellite throughout the 1960s starting with the basic CORONA camera on the left, through the stereo MURAL camera and the twin recovery capsule J-1 camera to the KH-4B J-3 camera with enhanced motion compensation on the right. Credit: Dwayne A. Day.

fuel to keep the orbit above the re-entry zone. But another reason was that they probably carried an additional payload either to pick up signals from sensors left by US spies inside the Soviet Union or to pick up Soviet electronic transmissions.

The phrase *Big Bird* is often associated with US spy satellites, but it wasn't until 1971 that the first of the satellites that bears that name was launched. The KH-9 Big Bird (official name: HEXAGON) weighed a staggering 13 tons and was the size of a Greyhound bus. Its primary purpose was to replace the 2-ton CORONA wide-area satellites, and it improved on the CORONA's performance first and most importantly by doubling the width of the swath or stripe of the ground covered in each photo—from the CORONA KH-4B's 290 km to 580 km. This extra width made it easier for the analysts to generate large-scale maps. Its resolution of 60 cm was also better than that of CORONA by a factor of at least 3.

So, this first of the Big Birds was six times the weight of CORONA to provide double its swath width and three times better resolution (i.e., $3 \times 2 = 6$). The arithmetic just balances out, but somehow you expect economies of scale to give you better than just the bare arithmetical product. Some people pointed this out at the time, and Big Bird was nearly canceled even quite late in its development.[69] Like CORONA it took stereo images, and it had double the number of return capsules (five instead of two, although the fifth was apparently always for the lower resolution—6 m— mapping camera[70]), had additional payloads to collect signals intelligence (radio and radar eavesdropping) and relay messages from covert ground sensors, and even carried an additional small satellite that it kicked out and left to continue monitoring

[69] Richelson (2002) p. 157.
[70] Day (2003) pp. 116–117.

Soviet radar signals after Big Bird had re-entered. As well as returning its film capsules every few days, Big Bird remained in orbit for 6 to 12 weeks to continue the work of these other payloads.

Eighteen of the monster Big Bird KH-9 satellites were launched between 1971 and 1986,[71] demonstrating that there was a strong demand for their imagery and other forms of intelligence.

The big disappointment with this original Big Bird was that it failed to address the great weakness of both CORONA and GAMBIT in not providing real-time imagery. We will see in Chapter 5 that the 1967 Arab–Israeli Six-Day War and the 1968 Soviet invasion of Czechoslovakia were over before US satellite imagery was in the hands of the analysts. In fact, the first few Big Birds did carry a real-time film read-out system to try and speed things up, but it seems not to have been a success and was discontinued.

Meanwhile, NASA and Soviet space probes were beaming beautiful images of the moon and the planets back to earth without the need for return capsules, so why couldn't the spy satellites do the same? The answer of course was that the planetary images had nothing like the pixel count of the spy satellite images, and the volume of the spy imagery was just too great for the technology of TV picture taking and radio return links. That is until the mid-1970s.

THE FIRST OF THE "MODERN" SATELLITES

The first of the spy satellites not to need film and to rely instead on digital camera technology was launched at the end of 1976. It was the last of the satellites that played a role in the negotiation of the SALT arms control treaties and thus is part of our story. At $13\frac{1}{2}$ tons it was slightly heavier than the KH-9 Big Bird and half again as long at 20 m. This satellite was called KH-11[72] and, although officially it is called KENNAN or perhaps CRYSTAL, most people also call it Big Bird, just like its film-carrying predecessor, perhaps because both of them fly overhead and are, well, big.

The civilian satellite industry was starting to use electro-optics (as the digital camera technology is called) in the early 1970s. NASA's civilian Earth Remote Sensing satellite, later called Landsat, was one, and Europe's Meteosat geostationary weather satellite, in which I played a small role, was another. The only surprise therefore was that it had taken the intelligence community so long to join the parade. A Presidential Science Advisory Committee panel chaired by Edwin Land (of Polaroid fame) in 1968 recommended that Big Bird be made digital and President Nixon was keen for it to happen during his term of office. But technical arguments delayed the decision to go into production and in 1971 Land took to personally lobbying key Congressional leaders and Administration officials. He is credited with persuading Nixon that it should go ahead and that his (Nixon's) backing was needed

[71] Wilson (1995) p. 270.
[72] Richelson (2002) pp. 198–201.

to overcome the bureaucrats. Nixon seems to have been persuaded more by Land's enthusiasm and vision than by his officials' views on costs and risks.[73]

The KH-11 Big Bird gets the images to the ground instantly by transmitting them to a special purpose satellite in a high orbit that relays them immediately to ground. Two of these Space Data System (SDS) satellites, whose name tells you nothing about what they do, were put in orbit shortly before the launch of the first KH-11. One of KH-11's limitations is that it requires more power than its solar panels generate to transmit all the data to SDS, so at least initially it was limited to doing so to two hours per day.[74]

Shortly after the launch of the first KH-11, its manufacturer, Lockheed, won the contract to build NASA's Hubble Space Telescope. It should come as no surprise therefore that Hubble and KH-11 look somewhat similar—Hubble points upwards and away from the earth, while KH-11 points downwards, but otherwise they do pretty much the same thing. Both have big telescopes with mirrors measuring about 2.4 m in diameter, so they can take pictures in poor lighting conditions—the bigger the mirror the more light it focuses into the camera shutter, hence the darker the objects that can be photographed. Another feature of Hubble that KH-11 probably mimics is the idea of having several cameras at the end of the telescope each with a particular capability—Figure 23. Thus, KH-11 might have a narrow-field high-resolution camera, a wide-area survey camera, a multi-spectral camera to analyse chemical constituents of the scene, and/or an infra-red camera for night-time vision.

The KH-11 stays in a higher orbit than earlier satellites—about 250×400 km—and that enables it to stay in orbit for much longer. Typically, a KH-11 remains in orbit for at least 3 years, and some have remained for more than 10 years. This shows the beauty of digital electro-optics as opposed to film, since the computer-like memory of the digital satellite can be used again and again once the images have been transmitted to the ground, whereas film can only be exposed once.

The resolution of KH-11 images has not been published, but presumably initially it was not as good as the GAMBIT KH-8 (8–10 cm) since a KH-8 was launched from time to time until 1984. A Big Bird KH-9 was also launched from time to time until 1986, so perhaps the swath width of the initial KH-11 Big Birds was not as wide as that of its capsule return predecessor.

Although the first KH-11 may not have produced images as sharp or wide as those of the best capsule return satellites, the second and later KH-11s gradually improved and by the early 1980s seem to have attained the same or better quality. The first KH-11 in 1976 used diode technology in its camera rather than a CCD, due to the still immature state of CCD technology. The second and subsequent flights used CCDs. The size of CCD dictates the quality of the pictures, and they got bigger as time went by (big is good). A CCD is an array of pixels and we know that the more pixels in a commercial digital camera the more you can enlarge the photos without graininess appearing. You perhaps recall that early consumer digital cameras with a large number of pixels were very expensive, but as the technology improved and the

[73] Temple (2005) pp. 443, 481–482.
[74] Richelson (2002) p. 201.

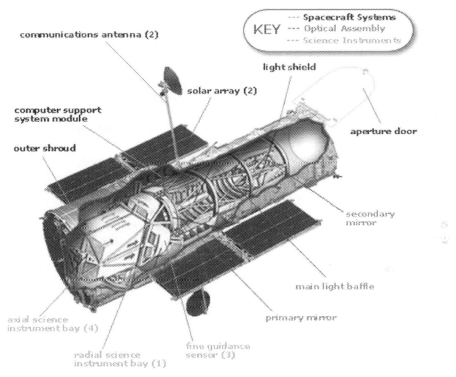

KEY
--- Spacecraft Systems
--- Optical Assembly
--- Science Instruments

communications antenna (2)

light shield

solar array (2)

computer support
system module

aperture door

outer shroud

secondary
mirror

main light baffle

axial science
instrument bay (4)

primary mirror

radial science
instrument bay (1)

fine guidance
sensor (3)

Figure 23. Schematic of the Hubble Space Telescope; four cameras can be inserted into the axial science instrument bay and one into the radial science instrument bay (at the left-hand end of the spacecraft) just behind the 2.4 m diameter primary mirror. Numbers in parentheses show how many of each item are on the satellite. Credit: NASA.

economies of scale kicked in—first 5 million pixels, then 8 million, and now 10 or more million—pixel cameras became affordable. The same trend occurred in the 1970s and 1980s with CCDs for professional applications.

The first CCDs available were not only very expensive they were relatively small, with fewer than 1 million pixels. In fact, at the beginning it wasn't clear that CCDs were better than TV-type technology for the most demanding imaging tasks. Work on the Hubble Space Telescope began in 1977 and it carried two cameras to take photos of the faintest and most distant objects—one supplied by the Jet Propulsion Laboratory (JPL) using the best available CCD, the other supplied by the European Space Agency (in which I was a team member) using so-called image intensification photon-counting technology—basically a souped-up TV camera with some clever built-in computer logic to reduce noise. A delay in launching Hubble due to the 1986 destruction of the Space Shuttle Challenger meant that the question had long since been resolved by the time it got into orbit in 1990—CCDs had improved by then to the point where they were much superior to the vacuum tube scanning electron gun of the TV technology. In any case neither of the cameras could show its true potential because Hubble's big mirror was warped and so both cameras produced blurred images.

So, we may surmise that each new Big Bird KH-11 incorporated a better CCD than the one before. Today the biggest CCD being manufactured has over 100 million pixels, and if that isn't enough for your camera, you can always use an array of CCDs—the Pan-STARRS camera in an observatory in Maui, Hawaii, has 1.4 billion pixels.[75] It takes a few years before satellites incorporate the latest ground-based technology because of the extra development needed to make that technology compatible with the vacuum and radiation in outer space, but in due course these giant CCDs will appear in a spy satellite of the future.

Satellites of the previous generation were launched occasionally after 1976, but the bulk of US satellite imagery now came from the Big Bird KH-11—of which two were usually in orbit at any given time, thus halving the time between over-flights of a target, and also enabling two different targets to be the focus of attention simultaneously. Between 1976 and 1988, eight KH-11s made it into orbit (one blew up on the launch pad in 1985) before the first of its successors, the 18-ton Advanced KH-11, was launched in 1992.

The image in Figure 24, apparently taken by a KH-11 Big Bird in 1984, was released to the media by Samuel Loring Morison, a US Naval intelligence analyst, resulting in his conviction and imprisonment—although he had served his two year prison term, he was pardoned by President Clinton in January 2001 in the latter's controversial final list of pardons. The image shows the Nikolaiev shipyard in Mykolayiv close to the Black Sea in what is now Ukraine. Under the giant gantry cranes we can make out the Kharkov Kiev-class nuclear-powered aircraft carrier under construction and what is apparently an amphibious landing craft behind it.

It's ironic that we know very little about the KH-11 because the Soviets actually know quite a bit about it. In 1977, not long after the first KH-11 was launched, a junior CIA officer, William Kampiles, sold the Soviets a copy of the 64-page KH-11 system technical manual for a paltry $3,000. Kampiles had only been with the CIA 8 months before being given a formal reprimand in writing for unsatisfactory performance, including sexual harassment of female colleagues. He resigned and when he left he stuck the office copy of the KH-11 manual in his pocket. Although the Soviets got the information, Kampiles was caught because the Soviet diplomat to whom he handed the document was a CIA spy. His eventual prison sentence of 40 years emphasizes the importance of the information in the manual. At the trial, CIA Deputy Director Leslie Dirks testified to its importance by listing some of the contents that would aid the Soviets, including illustrations of the quality of the images, limitations in its geographical coverage, and the process of photography it used, all of which could help the Soviets take action to prevent their facilities being observed, such as designing appropriate camouflage. Perhaps surprisingly, the Soviets seem not to have published the KH-11 manual, maybe because it would open them up to questions about their own satellites.

[75] Smart (2007).

Figure 24. 1984 KH-11 image of Soviet shipyard. Credit: National Security Archive.

By the way, in case you think I missed one of the Key Hole (KH) series out, the KH-10 was to have been the operational version of the Manned Orbiting Laboratory (MOL) which was canceled in the late 1960s. Others that I passed over include the KH-6 Lanyard, none of whose three 1963 flights was a success, and the KH-5 ARGON mapping satellite (550 km swath, 140 m resolution), 6 of whose 12 flights were successful in 1961–1964.[76]

SOVIET IMPROVEMENTS

Turning now to the Soviets, their systems initially avoided the need for separate wide area and close-look satellites by starting with mammoth satellites from day one. As noted near the start of this chapter the Zenit-2 was a 4.8-ton behemoth, able to carry both wide-area and close-look cameras on the same bird. While a CORONA return capsule weighed 88 kg, the Zenit one weighed 2.4 tons. After a dozen failed attempts, the first Zenit-2 images were returned successfully in August 1962, almost two years after the first successful CORONA images.

The Zenit-4 was even bigger at 5.5 tons, and from its introduction in late 1963 was launched in parallel to the Zenit-2 throughout the 1960s (Zenit-1 and Zenit-3 never existed). The main difference in the cameras between the two Zenit families

[76] Day *et al.* (1998) pp. 63, 74.

seems to have been that Zenit-2 had three small cameras lined up to give a 180 km swath, while Zenit-4 had a single larger camera that gave the same swath width on its own. The resolution of Zenit-4 is thought have been about 1 m, while that of Zenit-2 is said to have been in the 5–15 m range.[77]

Each Zenit satellite remained in orbit for about 8–12 days, and had only a single re-entry capsule—as explained earlier, the whole camera returned in the capsule not just the film. The launch rate of these satellites was extraordinary—a launch every 2 or 3 weeks, year after year. Of the 81 known (or assumed) Zenit-2 launches in the seven years the satellite was in use, 11 failed, 12 were partially successful, and 58 were totally successful. Success and failure rates for Zenit-4 have not been released, but at least 74 were launched.

The Soviet Yantar-2K system (sometimes called Phoenix or Feniks) in the 1970s and 1980s was the first radical upgrade to the Zenit system when introduced in late 1974. These $6\frac{1}{2}$-ton satellites took images with a resolution of about 50 cm and had three return capsules—the last one bringing back the camera as had the Zenit satellites. They had the same sort of maneuverability as the US Big Bird, able to lower their orbits to below 160 km, and stabilize there to observe a particular target area for several days, before raising the orbit again to reduce air drag. With the three-capsule feature they could "loiter" while awaiting a target that required their close-look capability, having sent back imagery from an earlier close-look in one of the capsules. They typically stayed in orbit for 4 or 5 weeks.

Six Yantar-2K were launched between 1975 and 1977 as part of a test program, two of which failed. Then in the 5 years 1978–1983, 30 more were launched, or about 1 every 2 months, of which 23 were successful and 3 partially so.[78]

Another enhancement to both Soviet and American missions was to use weather satellite information to plan the images. On average, half the images taken by the early satellites were obscured by cloud. To address this problem, in mid-1961 the US National Reconnaissance Office began a crash program to launch a weather satellite. The first successful launch of what eventually evolved into the Defense Meteorological Satellite Program was in June 1962. The results seem to have justified the cost because the program continues to this day, although statistics on the resulting proportion of cloud-free imagery have not been released. The military weather satellites have remained separate from the civilian satellites operated by the National Oceanographic and Atmospheric Administration (NOAA), at least in part because the ground terminals that pick up data from the satellites can be used to transmit "image" or "don't image" commands to the spy satellites.

A measure of the success of the military weather satellites is that the impetus for the military to send astronauts into space dried up. One of the objectives suggested for military spacemen was to check on the weather over the Soviet Union so that spy satellites could be targeted at the cloud-free areas. The Soviets too used weather satellites to schedule the imaging of its spy satellites, having also briefly toyed with using cosmonauts for this task.

---

[77] Day et al. (1998) pp. 157–170, and Gorin (1997) p. 441.
[78] Gorin (1998) pp. 309–320.

ACCURATE MAPS

The Soviets started out with enormous advantages over the Americans. Most of the Soviet Union was off limits to foreigners—and to most Soviet citizens—and maps could not be bought freely. The maps that were available pre-dated the communist revolution of 1917 and the first World War, and so were hopelessly out of date and inaccurate. American photo-analysts therefore had to build up maps from the images they collected. The location of Soviet cities and geographical features east of the Urals was often in error by 25 to 50 km,[79] and of course many places were not even known to exist. Between July and November 1941 (Hitler invaded the Soviet Union on June 22nd) Stalin ordered 1,500 factories to be transferred from the west to keep them out of German hands. One and a half million railway wagon loads carried the resulting cargo east—677 of the 1,500 to the Urals, 244 to western Siberia, and 308 to Kazakhstan and Central Asia.[80] Westerners had no access to these facilities and if their existence was known it was through garbled stories brought to the West by defectors. And these plants were the sort of military production facilities of considerable interest to the USA—the plants where missiles, bombs, and aircraft were manufactured.

By contrast, a Soviet citizen in the USA could buy detailed maps that showed, for example, the field boundaries that indicated the siting of a missile base.[81] They could then go and walk around the fence of that missile site. Local magazines and papers across America had stories about the military people and developments in their area, and these publications were of course available for anyone to purchase. Publications in the Soviet Union were all state-controlled and contained nothing but the officially approved *spin* on the news.

The early CORONA images were not designed to help create maps—they were there to count the number of missile bases, air fields, and submarines. By the mid-1960s the satellites had been enhanced with the addition of special sensors, better timing of each exposure, and better knowledge of the satellite's exact orbit, so that the error in assigning a location (latitude, longitude, height) to a point in an image in even the most remote part of the Soviet Union had improved to better than a mile.

The atmosphere limits what spy satellites can see in several ways. We have already noted that cloud cover prevented the early spy satellites from taking useful photographs much of the time. It would be the late 1980s before the first radar spy satellite able to see through the clouds was in orbit—more about this later (Chapter 9). Even when the sky is clear of clouds, the hours of darkness prevent detailed photographs being taken. Most of the Soviet Union is north of 50° latitude and much of it above 60°. The good news for American spy satellites was that the summer provided long hours of daylight, but the bad news was that winter nights were equally long. Paranoid American military analysts worried about what the Soviets might be up to during those long hours of darkness. By contrast, all of the continental USA is

[79] Day *et al.* (1998) p. 210.
[80] Werth (1964) pp. 209–218.
[81] Day *et al.* (1998) p. 210.

south of the 49th parallel, making the summer/winter contrast less extreme for Soviet spies in the sky.

Even when clear and in daytime, the atmosphere is turbulent. We see this for ourselves when viewing the stars—the twinkling of starlight is due to the shimmering of the atmosphere. This atmospheric turbulence puts limits on the accuracy of spy satellites. Although accuracy figures for the latest spy satellites are secret, we can work out the accuracy achievable with the Hubble Space Telescope if it were to point at the earth, which will give us a rough idea of what is possible. Hubble was designed to have a resolution of about 30 millionths of a degree. If the atmosphere were completely still and Hubble was pointing at the earth from a typical spy satellite altitude of 250 kilometers it could resolve objects on the earth's surface of about 10 cm (4 inches) in size.

What do we mean by a resolution of 10 cm? It doesn't mean that two objects 10 cm apart will be recognized as two objects and that two objects 8 cm apart will be recognized as a single larger object. The ability to detect the small gap between two objects will depend on factors such as lighting conditions, the shapes and surfaces of the objects, shadowing in the gap, and the color and sheen contrast between the two objects and between them and the gap. In general, objects 10 cm apart will be recognized as being separate, while objects 8 cm apart will be more likely to be recognized as a single object, and a gap of 1 cm between two objects would hardly ever be detected.

Telescopes can focus an image onto a film down to a limit set by the wave-length or color of the light. The resolution of the spy satellites comes down to how sharply you can focus light onto the film and the inherent graininess of the film itself. Early spy satellites generally worked at the limit of both of those parameters—the smallest image picked up by the telescope being roughly equal to the graininess of the film.

Digital cameras and the camera in our cell or mobile phone have familiarized us with the term *pixel* or picture element. A camera that takes images containing 5 million pixels gives better pictures than one that contains 2 million pixels. If you blow up the picture the graininess in the 2 million pixel image becomes evident, whereas the 5 million pixel image still looks sharp. When graininess appears, you have reached the ultimate resolution of the picture—objects smaller than the graininess are blurred and can't be resolved.

Even if a spy satellite has a resolution of, say, 50 cm, it will be possible to interpret features smaller than that. As we will see later, it is desirable to be able to work out the width of a missile to see whether it complies with a treaty agreement. It should be possible to tell this with an accuracy as much as 10 times better than the resolution, because each edge of the missile is made up of several pixels, giving us an averaging effect. I can speak for this personally through two commercial systems that I have been involved with at LogicaCMG in recent years. In the first one, for a Japanese customer, our software compares features in a satellite picture of the earth with features stored in a computerized digital map. The purpose is to work out from this comparison how much the image is distorted or the satellite is mis-pointing. The resulting accuracy is 10 times better than the size of an individual pixel. The second system was for a European customer and achieved similar sub-pixel accuracy in monitoring the movement of clouds from one satellite image to the next—thereby measuring the speed and direction of the wind.

So, the quality of a spy satellite image is not a simple resolution value in meters or centimeters. A sophisticated scale of quality from zero (worst) to nine (best) was defined by the National Reconnaissance Office in terms of the information you could obtain from the image. For example, to say that an image was level 4 meant that you could see whether the door of a missile silo was open or closed and was equivalent to a resolution of about 2 meters. At level 6 or about 50 cm resolution you could distinguish between several different types of missile.

Robert Kohler, then at the CIA, recalls that in addition to defining what each quality level was in words, they tried to have an image that illustrated each level. It proved difficult to find an image, even a low-level airborne image, with level 9 quality—the highest possible quality level. Finally, a picture taken by an aircraft flying along the border between East and West Germany proved to have the required quality—it showed an East German soldier urinating. The image was displayed for all to see under the banner headline "German soldier pissing in the snow—level 9."[82]

The four pictures in Figures 25a, 25b, 25c, and 25d are courtesy of John Pike and his GlobalSecurity organization.[83] They show two newspaper front pages, a vehicle licence plate, and a golf ball in images of various resolutions. The first image with a resolution of 10 cm is probably the best that spy satellites can do, and doesn't come close to reading the headlines or the number plate nor even detect the golf ball—no chance to use these satellites to find that golf ball in the rough then. The second image shows the quality available from some aerial imaging systems, and this probably can detect the golf ball. The third image with a resolution of 1 cm allows you to read large tabloid headlines, but not the licence plate or normal headlines—but interestingly does resolve the picture on the *New York Times* front page quite well. The final image with 1 mm resolution shows what the other images miss—the golf ball is distinguished as circular and not a small white box, the licence plate is revealed, and headlines, pictures, and text in both newspapers are legible.

The Hubble Space Telescope was developed more than 20 years ago, so we might expect current spy satellites to be somewhat better in their performance. But the fact remains that resolving from space details on the surface of the earth that are smaller than a few centimeters is almost impossible. Forget about recognizing Osama Bin Laden as he walks down the street (even if he conveniently looks up to the sky at the right moment) or reading car number plates—or the newspaper headline. Photos that have been released show that it might occasionally be possible to identify the make and model of a car, although probably not its color—color cameras in space tend to have worse resolution than black and white.

And all of that assumes that the atmosphere is absolutely still— which it hardly ever is. The atmosphere is constantly moving due to thermal gradients within it. On a really hot day the shimmering is visible to the naked eye—for example, above a tarmac surface or a hot sandy beach. As you look further and further through the

---

[82] McDonald (2002) p. 223.
[83] *http://www.globalsecurity.org*

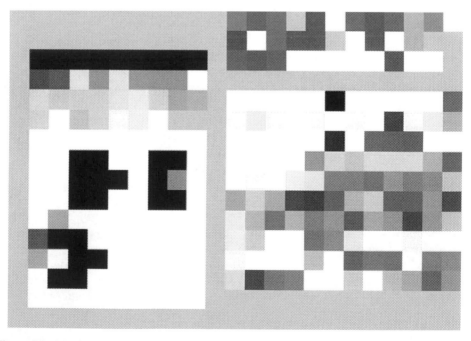

Figure 25. (a) 10 cm resolution (high-resolution spy satellite); (b) 3 cm resolution (high-quality aerial photo). Photos courtesy: John Pike and *GlobalSecurity.org*

Figure 25. (c) 1 cm resolution; (d) 1 mm resolution. Photos courtesy: John Pike and *GlobalSecurity.org*

air the thermal shimmering accumulates, so that through a telescope or strong binoculars the shimmer is quite apparent. Even in the cool of the night, the shimmering air causes the stars to twinkle. In astronomical photos, the atmospheric shimmer is about one second of arc, which in a photo of an object 100 km distant is about 50 cm. Thus because of this shimmer, an astronomical telescope photographing a satellite overhead could not make out features smaller than 50 cm—and more or less the same applies in the reverse direction with a spy satellite taking photos of the ground.

In the past 10 years, astronomers have found a way to counteract the constant shimmer in the air. They monitor a star that is so distant that it can be considered as a point, then watch how the shimmer in the air makes the star move its position. It turns out that the star will be unmoving for a millisecond or so between shimmers. So by monitoring this movement every millisecond (needless to say, computers do the monitoring, not humans) the movement of the air can be determined—that is to say, the movement of the air in the same part of the sky as the star being monitored. This works well for astronomers who just have to choose a suitable distant star in the part of the sky they are observing.[84] The movement of the guide star is used to adjust a mirror within the telescope to compensate for the atmospheric shimmer.

However, this technique doesn't work so well in the opposite direction (i.e., for a spy satellite looking down on the earth from above the atmosphere). For one thing the satellite is moving at 29,000 km/h (8 km/s) across the scene of interest. To get a feeling for that speed, the satellite crosses a 20 meter street, say, in $2\frac{1}{2}$ milliseconds. So, every few milliseconds the spy satellite would need to find the equivalent of a new guide star to act as a reference for calculating the shimmer. And anyway there probably isn't anything conveniently equivalent to a guide star. To take advantage of the method used by astronomers, a spy satellite would have to analyse each scene and find an object it could be sure was a point or perhaps a thin line, then watch how that point blurred and introduce compensations to the camera.

"My digital camera does this" I hear you say. And of course it is true that the technology for ensuring blur-free images from modern digital cameras and camcorders might well be able to remove some of the atmospheric shimmer from spy satellite images. If that is so, nobody is talking about it.

In general, the effect of atmospheric shimmer looking down from space is less than when looking upwards at the stars.[85] Most of the shimmer is caused in the lower atmosphere where the air is most dense. It's a bit like looking at an object through a smoked glass pane. If you stand up close to the pane and look at distant objects the result is very blurred. On the other hand if you look at an object that is just on the other side of the smoked glass and you are standing back from it, you can make out quite a bit of detail. The part of the atmosphere that is shimmering is close to the earth's surface, so a spy satellite is "standing well back" thus limiting the shimmer

---

[84] Actually, the main problem for astronomers is that the reference stars, or *guide stars* as they call them, have to be fairly bright so that they can be detected rapidly and accurately, and there are large parts of the sky without suitable bright stars.

[85] Day *et al.* (1998) p. 36.

effect. An astronomer looking upwards is close to the shimmering air and trying to look at objects far away.

The films returned by Soviet and American satellites had to be magnified hundreds of times to reveal detailed intelligence information. One of the concerns therefore was to avoid even a speck of dust entering into the processing and printing process. Since much of the information was being obtained at close to the resolution of the films, a smear or a dust spot could be mistaken for a physical object of significant size. Special film processing and printing laboratories had to be created to ensure that the quality of the material was maintained at all time.

CORONA, Zenit, and their direct successors took great pictures during the day under clear skies, but in cloudy conditions and at night they were blind. Nowadays even clouds and darkness don't prevent some satellites from spotting what's happening—we will return to this topic in Chapter 9. However, first let's look at what spy satellites *can't* do, but Hollywood would have you believe that they can.

THE HOLLYWOOD VERSION

Movies and TV shows frequently depict a spy satellite following a scene as it unfolds on the ground. *Mission Impossible II*, for example, has a spy satellite hover over Sydney while intelligence agents watch the bad guy below. Satellites like that exist only in the movies, although Hollywood intends the viewer to treat it as technically accurate—not science fiction as in "Beam me up, Scotty" from Star Trek. Spy satellites take still pictures not moving pictures, in black and white, and at most can fix on the same target for a few seconds. They can make out objects of about 5 cm in size and as discussed above can't read car number plates as depicted in the 1997 movie *Shadow Conspiracy* starring Charlie Sheen.

Another Hollywood invention is that satellites can be diverted from their regular tasks at the bidding of a junior analyst. In a 2006 episode of TV's NCIS series, two "teccies" at an agency vaguely like the CIA are using a spy satellite to watch sunbathing girls when they accidentally spot a murder. The more proactive of the teccies is 20 years old and just 6 months out of university. In practice of course, control of these billion dollar satellites is much more tightly and bureaucratically overseen than this.

The 1998 conspiracy movie *Enemy of the State* with Will Smith and Gene Hackman makes the satellites the star of the show, following the heroes through every twist and turn. On the other hand, the 1992 movie of Tom Clancy's *Patriot Games* starring Harrison Ford is reasonably honest in its use of spy satellite images, showing only blurred and grainy pictures, and recognizing that terrorists know the schedule for spy satellites to be passing overhead and can hide themselves and camouflage equipment at the scheduled times. Which of these movies did I enjoy the most? I have to tell you that *Enemy of the State* won hands down—you can't beat a good government conspiracy plot with loads of hi-tech gadgetry, no matter how absurd its technical detail.

5

Problems of verifying an Arms Limitation Treaty

Having explored what imaging satellites could do in the last chapter, let's look at how those images can be turned into information for verifying arms treaties. Satellites are not the only source of information used for this purpose, so the radars and eavesdropping facilities that complement the satellites are briefly described. Then we discuss what the early satellite flights found—and what they couldn't find. They may have prevented nuclear war, but short, sharp engagements like the Arab–Israeli Six-Day War are some of the examples used to illustrate their limitations. Finally, we come to the *dark side* of spy satellites—they helped create political stability, but they were also a fountain of information for choosing the targets for the nuclear bombers and missiles that threatened that stability. Let's start with some history.

The mutual suspicion between the Soviet Union and the USA led to both sides requiring some form of verification of any agreement. Suggestions by either country to allow on-site inspections of the other side's facilities were interpreted as attempted spying. The route to an agreement therefore included long negotiations on what forms of verification would be permitted in addition to the extremely difficult negotiations on the main topic of limiting missiles, bombs, planes, submarines, etc.

The first significant agreement came in 1958, but instead of a negotiated Treaty, it comprised unilateral statements by both countries to stop testing nuclear weapons in the atmosphere. There were no formal negotiations leading up to this important arms agreement, and it was not written down and signed. Following contacts at ambassadorial level, President Eisenhower and Premier Khrushchev simply and separately pledged their governments to cease such testing if the other side did too.[86]

The primary American reason for agreeing to halt atmospheric testing was the public outcry over the exposure of fishermen to the effects of radiation from the American hydrogen bomb tests in the Pacific Ocean. In addition, Eisenhower seems personally to have wanted to take some action in the direction of stopping the

[86] York (1983).

seemingly inexorable nuclear escalation between the two super-powers. Premier Khrushchev, too, seems to have been similarly motivated.

Both sides began instead to test their nuclear bombs underground, thus avoiding radioactive fallout.

The fragility of this unsigned agreement became apparent barely 2 years later when in 1960 Eisenhower declared the "moratorium" on atmospheric testing over, although adding that the USA would give notice before any tests were actually undertaken. The Americans had become disillusioned by the glacial progress being made in negotiations on a Treaty to formalize the test ban. In early 1960 France conducted a nuclear test in the atmosphere, thus giving the Soviets the excuse to resume above-ground testing.

The Soviets didn't immediately break the moratorium. But when they did, in 1961, their insensitivity to the effects of radiation on the populace was abundantly clear. Despite the opening up of Soviet archives since the end of the Cold War, we do not yet have a clear picture of the damage to life and ecology caused by the massive Soviet tests that followed. Large areas around the testing grounds in Kazakhstan and near the manufacturing facilities in the Chelyabinsk region are still out of bounds to visitors for security reasons, but the damage seems to have been on a scale similar to that of the Chernobyl civilian accident 20 years later in 1986—see Chapter 6.

HOW TO VERIFY

Three main forms of verification of any agreement were available to both the Soviets and the USA. The first was ground-based radars to detect the launch of a missile. The detection could be achieved over long distances, due to the use of so-called over-the-horizon radar. The second was radio-listening stations on land, ship, aircraft, and even satellite that could monitor the radio transmissions between a missile and the ground. This *telemetry* information is used by the country undertaking the tests to alert the ground controllers as to the status of the missile—detection of unintended echoes from the moon's surface by giant antennas on earth were also used.[87] The third was satellite images primarily to count the number of deployed missiles and aircraft and their nuclear payloads, and monitor their manufacture and transportation, and also to detect the launch of a missile by its tell-tale bright rocket plume.

The overall approach to verification eventually worked out between the super-powers in SALT-I and SALT-II relied on the fact that missiles had first to be tested several times before they entered production and then deployment. The testing of missiles could be followed in considerable detail by the other side using ground radars, radio-listening ships and the other means mentioned above. Test ranges were built by both countries instrumented with their own radars largely over ocean regions to avoid failed missiles landing on populated areas. The ocean test ranges were open to ships from any country, so the other side knew where to deploy its ships in order to follow the results of the tests. Testing of new missiles is still an essential part of their

---

[87] Richelson (2002) p. 89.

development, although the number of test flights is perhaps coming down as computer simulation techniques and specialized test facilities improve—for example, the French M45 missile introduced in the 1990s needed 34 test flights before becoming operational, whereas its successor the M51 will probably require fewer than 10.[88]

Many of the Soviet missile tests ended up in the northern Pacific Ocean, watched by several US radar stations in the region. US soil itself, in the shape of Smeya Island far out in the Aleutian chain, was home to one such radar station. Situated so far to the west that the International Date Line does a dog leg to the left at that point to keep its day the same as for the rest of the USA, Smeya is geographically a lot closer to mainland Asia (750 km) than to mainland America (2,000 km). Thousands of kilometers to the south, stations on Midway and Johnston Islands act as a radar picket for planes and missiles approaching the Hawaiian Islands—Midway about 2,000 km to the northwest, Johnston 1,200 km to the southeast. These installations plus radars 3,000 km to the west on the Bikini and Kwajalein Atolls in the Marshall Islands can pick up missiles farther south in the Pacific. Next, high-power radars in Japan's northerly Hokkaido Island, in Taiwan, and in the Philippines use the over-the-horizon (OTH) mode of tracking to watch missiles deep inside the Soviet Union. OTH means that the radars in the Pacific transmit signals that bounce off the missiles and their reflections are picked up in Europe. Finally, radars in Turkey pick up the early phase of missiles from some of the Soviet launch sites.

The key phases of a missile's flight were at the start and the end. Being able to monitor the launch and first few minutes of flight of a missile gave important information about its power and range, but unfortunately US radars did not have a clear line of sight to the Soviet launch sites, and so generally missed this phase. Having flown high into space, the missile's re-entry into the atmosphere was equally critical. US radars monitored whether the missile could alter its trajectory during the re-entry to compensate for guidance errors during the flight. They could also observe how many warheads re-entered, and if more than one, whether *they* could maneuver (Figure 26). A missile with three warheads each of which could maneuver was as good as three separate missiles—they could be targeted at specific points tens of kilometers apart. By comparison, a missile with three warheads that landed without maneuvering was not much better than a single very large missile—all three warheads would land within a few kilometers of each other.

The OTH radars could monitor not only the trajectory of the missiles but also the disturbances in the ionosphere caused by the jet of gas emitted by the rocket. It turns out that the pattern of that disturbance is distinct for each type of missile—a sort of signature for a particular type of missile.[89]

When it came to intercepting radio transmissions, the second of the three technologies of verification, the USA had some advantages in this listening game. Its allies ringed the Soviet Union, and their fortuitous geographical location was put to good use. The Norwegians, for example, were funded to install listening stations close to the Soviet border, and thus well placed to monitor submarines, missiles, and the

[88] Dupont (2006).
[89] Greenwood (1973).

Figure 26. Time lapse photo of the re-entry of the eight MIRV warheads of a Peacekeeper ICBM.

associated radars and communications in one of only two main Soviet submarine bases—the other based in the Crimea far to the south. Norwegian boats, too, were outfitted with US electronic equipment to help pick up signals intelligence far out in the northern seas.

Another useful location was Iran which was as close as the US could get to the Soviet missile and satellite launch complex at Tyuratam—1,000 km to the northeast. The equipment used to monitor the distant Soviet activities was sophisticated but living conditions for the handful of CIA operatives manning the station were not. The site was in a remote and isolated mountainous region to which supplies had to be flown in from Tehran and even water had to be carried up the mountain. Although the staff could console themselves that it was like camping out for the year-long assignment there, the latrine facilities left something to be desired—basically just a slit trench, which in the frequently freezing weather was extremely unpleasant.

Iran continued to be a mainstay of US electronics eavesdropping until January 1979 when the Shah of Iran was deposed and Ayatollah Khomeini's supporters swept to power. The equipment was left running by the Iranians as they wanted to show it to potential buyers and if they switched it off they were not sure they could get it running again.

Eight years earlier the then US national security adviser (later Secretary of State) Henry Kissinger set in motion a series of events that would create a surprising alternative to Iran. He traveled to Beijing in great secrecy to discuss the establishment of diplomatic relations between the USA and communist China, and during that historic visit he offered to give the Chinese US spy satellite images and electronic intercepts of Soviet forces along the China–Soviet border, an offer that the Chinese accepted. By late 1979 discussions on joint spying on the Soviets had been agreed in principle and by 1981 two stations were in operation in the far northwest of China.[90] The Qitai station's location was north of the Tien Shan mountains with their peaks climbing to over 7,000 m and thus had a relatively clear view west and northwest. This in particular provided visibility of telemetry data from the early phases of missile launches from the Tyuratam (Baikonur) launch site a bit less than 2,000 km away in Kazakhstan—early-phase telemetry was particularly useful in characterizing the missile's rocket engine. The anti-ballistic missile radar site at Sary Shagan to the west of Lake Balksah in Kazakhstan that had caused so much heartache for the SALT negotiators (as we will see in Chapters 6 and 7) was also within easy listening distance.

The second station was about 200 km to the southwest at Korla on the southern side of the Tien Shan mountains. From this location it had visibility to the northeast towards the Kamchatka peninsula—the other end of the Soviet missile test range. Both stations were close to one of China's most peculiar geographical features, the Turpan Depression, which despite its proximity to the towering and perennially snow-capped Tien Shan mountains is the lowest and hottest place in China— 150 m below sea level and temperatures exceeding 43°C (110°F).

Australia's geographical location thousands of kilometers from the Soviet borders seemed an unpromising location for a radio-listening station. But Australia's stance as a staunch ally of the US (sending troops to Vietnam when Britain, for example, refused) proved surprisingly useful in monitoring Soviet activities. In the second half of the 1960s the USA conceived the idea of using a geostationary satellite to listen in to Soviet missile telemetry. Such satellites can view almost the whole Soviet Union continuously from their location 35,000 km out in space—at which location Sir Arthur C. Clarke had pointed out in 1947 they would appear stationary from the ground. The satellite then transmitted what it had received down to a station on the ground, which is where Australia came in. By placing the station at Pine Gap in the center of Australia, close to Alice Springs, the transmissions were kept far from the prying ears of, and potential electronic interference from, Soviet ships and planes.

Not everyone was a fan of the Australian locations. Another US facility was built at Nurrungar close to Woomera in South Australia to receive signals from the Defense Support Program (DSP) launch detection satellites. Although a lot closer to the coast than Alice Springs, the Woomera site was "at the end of the world" according to US Air Force Under Secretary John McLucas after his first (and last) trip there.[91]

[90] Richelson (2002) pp. 217–218.
[91] Richelson (1999) p. 54.

There was initial skepticism as to whether a geostationary satellite could separate out useful signals from the loads of mundane and unwanted radio traffic emanating from the area under surveillance—it was felt, for example, that TV signals would drown out the missile telemetry signals. Detailed technical calculations showed that it *was* possible to pick out the wanted signals from the mass of background transmissions, and so the RHYOLITE and CANYON programs got underway—CIA and US Air Force programs, respectively. Future US Defense Secretary William Perry is credited with establishing the workability of the concept (he was an electronics engineer at the time), and CIA official Lloyd Lauderdale proposed the mechanism for unfurling in space the gigantic antenna needed by these satellites—he apparently brought a French umbrella to work to show how it could be done.[92]

The first CANYON satellite was launched in August 1968 with the primary mission of intercepting Soviet bloc radio communication transmissions, but with the ability it turned out to also pick up missile telemetry signals. The first of the CIA's RHYOLITE satellites followed in June 1969 intended specifically to listen in to missile telemetry transmissions.

Getting useful information from the telemetry data of a missile someone else has built was not easy. At first glance the data seemed random, but after some analysis patterns emerged. When the engine in the missile shut off, various parameters in the telemetry changed—acceleration would stop, gas pressures inside the fuel tanks would fall, temperatures in the nozzle would reduce, and so on. National Security Archivist Jeffrey T. Richelson compares the problem to having a dashboard in a new car with no writing on the dials. After a while it would become clear which dials referred to speed, which to RPM, oil pressure, water temperature, etc. Even the units of measurement for the various parameters could be estimated provided you have enough data over a range of conditions.[93]

The surreptitious listening stations that monitored missile launches could also characterize the radars used by the adversary. By knowing the technical details of a radar the adversary could figure out ways to fool it. The technical characteristics of the radar also revealed its purpose. America spent several years trying to decide on the purpose of one gigantic radar in Kazakhstan. First spotted in U-2 and CORONA images in 1960, one of the radars was a colossal 200 m long, using a technology called a phased array antenna rather than a conventional dish. With an enormous 25 MW transmitter this radar was one of the world's most powerful and many analysts felt that its role must be as part of a defensive system to shoot down incoming missiles— anti-ballistic missile (ABM) as the technique was called. It wasn't until 1965 that a large CIA antenna in California was able to pick up echoes of this Soviet radar that bounced off the moon's surface and thereby get enough information to work out its capability. The Soviet radar was indeed intended for ABM tests and could also track satellites and missiles.

In Chapter 4 we mentioned the GRAB electronic listening satellite, and it's not clear why that satellite and its successors were not able to answer this question

---

[92] Richelson (2002) p. 111.
[93] Richelson (2002) p. 86.

sooner. Perhaps the Soviets deliberately avoided using the radar when that satellite was in view—it only came over the Soviet Union a few times a day for less than half an hour at a time.

New missiles were initially very unreliable, and the early ones required 20 or 30 test launches before they were sufficiently reliable to enter production. The listening ships and stations of the other side therefore had ample opportunities to characterize a new missile before it entered production in terms of, for example, its size, range, acceleration, trajectory, descent profile, etc. Even the color and brightness of its exhaust plumes were useful information, indicating the type of fuel used.

Despite the plethora of radar stations and eavesdropping facilities and satellites, the US lacked a means to reliably detect operational missile launches as they happened. The facilities described above were great for following missiles that originated from the major test sites, but were not always able to pick up launches from the hundreds of missile silos across the Soviet Union and submarines at sea. Early-warning satellites were eventually launched to perform this function with the deliberately uninformative title of the Defense Support Program (DSP) satellites. They were designed to watch out for an enemy launching a missile at the USA in anger, not to assist in Treaty verification. But, of course, they could detect the launch of test missiles and provide useful information to the verification analysts.

The first DSP was launched in November 1970, although a series of experimental satellites, with the designations MIDAS and RTS, had been launched throughout the 1960–1966 period to demonstrate the concept of launch detection and try out various techniques—different orbits, sensors, data recovery methods and so on. The general idea was to watch for the bright flash of light and flare of heat given off by a rocket motor. They proved very successful at spotting Soviet and Chinese missile launches—by the end of June 1973 the DSP satellites had detected 1,014 launches, and they added another 982 in the ensuing 18 months. This vast quantity of data provided useful information to the teams of analysts trying to piece together the missile development and deployment plans of the Soviet military. DSP data allowed the launch site to be pinpointed to within 3 to 15 km and the launch heading to within 5° to 25° depending on various factors such as the relative location of the launch and the DSP satellite. There were a few false alarms, but only for smaller missiles (no false reports of ICBM launches were lodged), such as submarine-launched missiles in the northern hemisphere summer (due to sun/sea glint).

Any bright flash triggered the DSP sensors, but software algorithms sorted out the missile launches from ammunition dump explosions, forest fires, gas pipeline fires, the burn-up of satellites re-entering the earth's atmosphere, military jet aircraft using afterburners—even the July 1996 explosion of TWA Flight 800.

In addition to sensors to detect the bright trace of a rocket launch, DSP satellites carried special instruments to spot nuclear explosions, taking over the role initially performed by the US VELA satellites dedicated to nuclear explosion detection. The specialized instruments on VELA and DSP could detect X-rays, gamma-rays, and neutrons given of by an above-ground nuclear explosion, and did so very reliably. The one exception was the event recorded by a VELA satellite in the early hours of September 22nd 1979 about 2,500 km off the coast of South Africa. The detectors on the two DSP satellites that had a view of that region failed to spot the event, and aerial sampling missions by the Air Force and laboratory analysis of foliage from southern Africa failed to resolve the matter. Whether this was a South African and/or Israeli nuclear test remains unresolved to this day.[94]

The third verification technology, satellite imagery, was important during the test phase. The imagery enabled the analysts to monitor where the missiles had been built by spotting the movement of trains and other forms of large transportation, thereby identifying the factories in which the new missile was being developed, and the characteristics of its transportation. Image-interpreters claimed to be able to tell from the shape and size of the containers which missile was contained within, and this information was built up laboriously by linking missile tests to their containers.

Each form of missile had its own arrangements for command and control—to fuel the missile, arm it ready for launch, monitor its status, etc. As the 1960s progressed, missiles increasingly were based in underground silos, providing some protection against surprise attack—although not against nuclear attack. The size of the silo, and the arrangement of buildings, radars, anti-aircraft missile batteries, access roads, perimeter fencing, and other civil works was typical and regular for each type of missile, especially in the centrally managed Soviet economy. Thus, American image interpreters could identify the type of missiles at a site from its geographical layout.

Detection of a new Soviet missile emplacement often started with spotting new roads or a new rail spur. The Soviet Union has vast areas with almost no population, so activity in these areas stood out. The rigid Soviet planning system engendered a uniformity of style and architecture in every type of building. Local constraints such as existing houses and farm boundaries meant nothing. This led to a lack of variation in the construction layout of the various missile bases which helped US analysts to characterize them.

Foreigners were forbidden from traveling in the Soviet Union unless escorted by an official guide. Most of the country was out of bounds to visitors—escorted or not. There was no free press, just newspapers, radio, and TV owned and controlled by the government. So, Western analysts had to rely on irregular and unreliable sources of information such as the occasional defector—and as former UN Weapons Inspector chief Hans Blix has pointed out in connection with recent nuclear intelligence controversies, "defectors didn't want inspection, they wanted invasion."[95] The concept of a defector was also peculiar to the period, and existed because Soviet citizens were forbidden to leave their own country. The Berlin wall was just the most visible manifestation of this restriction—armed guards patrolled the border of the Soviet-controlled world with orders to shoot to kill. Winston Churchill didn't realize how literally true were his words when he spoke in 1946 of "an iron curtain descending across the continent of Europe from Stettin in the Baltic to Trieste in the Adriatic." The phrase "iron curtain" came to symbolize the prison created by the Soviets for their own citizens and the citizens of subject states, such as those in eastern Europe.

---

[94] Richelson (1999) pp. 74, 78–81, 104–105, 108–109.
[95] Speaking on *Sky News*, 12 March 2007.

THE BOMBER AND MISSILE GAPS

Western knowledge of Soviet military capabilities was very sketchy. A US diplomat invited to watch the annual Red Air Force Day air parade near Moscow in 1955 reported that the number of Bison strategic bombers in the fly-by was more than double the number in the May Day parade fly-by earlier that year. The information was passed to Washington, causing the US to raise its future forecast of the Soviet Bison force. In fact, the Soviets had deliberately hood-winked the Western observers by having most of the planes fly past twice, thus appearing to inflate the size of the bomber fleet. The 18 bombers that had actually flown had been inflated by this device to about 30, representing a significant increase compared with earlier that year. The Americans wrongly attributed the increase to rapid production of new Bisons, and extrapolated that production rate to forecast 600–800 Bisons by 1960.[96] In later years, of course, satellites would have counted the bombers on the ground and not been fooled this way.

The implications of the apparent increase in Soviet bomber numbers were significant. The US Air Force immediately called for a similar increase in American B-52 bombers, which a reluctant President Eisenhower agreed to. Later when the Soviet deceit became evident, Eisenhower resolved to avoid allowing the US military to be his main source of military intelligence, and to give that responsibility to the CIA instead. He saw only too clearly the vested interest in the military to use inflated estimates of Soviet military power to buttress the arguments for increases in US military forces.[97] He saw to it that it was the CIA, therefore, that managed the U-2 flights, and later the first series of spy satellites. As CIA Deputy Director Ray Cline put it "A cardinal rule of intelligence prohibits the same unit from conducting intelligence operations and then having the right of exclusive evaluation of the results."[98]

The Soviet motivation in mounting the deception was to enhance their negotiating position at the imminent Summit meeting in Geneva with President Eisenhower and other Western leaders. Ironically, the long-term effect was to worsen the Soviet military position, since within a few years the American B-52 bomber force vastly out-numbered the relatively small Soviet force.

Before the spy satellite, the U-2 aircraft was therefore the only reliable source of information available to the Americans on what was happening across the vast reaches of the Soviet Union. Despite the risk of the U-2 being shot down, President Eisenhower thought it worthwhile to authorize U-2 flights from time to time.

Satellites could determine the number and type of missiles and aircraft deployed. They could also detect the number of submarines being built since their construction could only take place at two Soviet shipyards and each one took years to complete (Figure 27). The ability to detect submarines once they were at sea and underwater is still not admitted in detail by either side, but the US certainly had extensive

---

[96] Burrows (1986) p. 68.
[97] Ambrose (1984) pp. 476–477.
[98] Burrows (1986) p. 216.

Figure 27. June 9th 1966, Gambit KH-7 image of submarine pens in Polyarny shipyard, Murmansk fjord (68°N, 33°E). Credit: US Geological Survey.

submarine detection facilities strung across the entrances to the North Atlantic between Greenland and Russia.[99] To introduce a new type of missile would require many test flights of that missile, so the ability to monitor the data from each missile launch was important. In later Treaty negotiations, both sides accepted that if you could monitor missile launches you could monitor whether new missile types were being developed.

There were still elements of an arms limit that were difficult to monitor—that is to say, infringements might be possible without detection. For example, the same number of missiles could hit several times that number of targets if the missiles were converted from single warheads to MIRVs. A MIRV missile requires several re-entry shields, guidance systems, etc., so 10 MIRVs are a lot heavier than one warhead and

[99] Garwin (1972).

therefore require a much heavier and larger missile. By preventing the introduction of a new missile type or the radical improvement of existing ones it was considered that this issue could be monitored.

Cruise missiles were another difficult topic. These are aircraft-like vehicles (without a pilot) that can be launched from the ground, an aircraft, a ship, or a submarine. They were the direct descendants of the V1 buzz bombs that Germany rained on England in World War II. The Cold War variants, however, could carry nuclear weapons. It was eventually agreed to put an upper limit on the range of these vehicles (600 km in SALT-II) and to monitor that limit primarily by not permitting any tests that were greater than that range. It was argued that a country would need to test a weapon of this type at its full range if it was to rely on it, so by forbidding the testing you effectively forbad the weapon itself. However, cruise missiles are not like rockets. A rocket's full range depends on the power of the rocket motor and that has to be tested in order to be sure it will achieve its goals. A cruise missile's range depends on how much fuel you put in it not on the power of the engine. So, you can test the navigation accuracy of a cruise missile at relatively short range, then fill the fuel tank and confidently expect it to work at long range. This issue was considered insurmountable in the 1970s, and as a result cruise missiles were recognized as an area of SALT-II that was effectively only partially verifiable.

An aircraft could carry a large number of cruise missiles and it was difficult to see how to verify those numbers. The total number of cruise missiles could be broadly monitored and the total number of long-range aircraft capable of carrying them could also be monitored. So, in the end the agreement limited those two numbers and avoided trying to limit the number per aircraft. The US was much more advanced than the Soviets in the miniaturization needed to build airborne or submarine-borne cruise missiles, so the agreement effectively favored the USA, at least initially.

Submarine-launched missiles were also difficult to monitor accurately. The total number of long-range submarines (i.e., with nuclear-powered engines) could be verified—satellites could watch them being built and/or launched. And the missile capacity of such submarines was known or at least estimated. There remained only the problem of knowing whether the missiles carried MIRVs or not, as already discussed above.

These days we are used to systems that monitor images automatically. The congestion-charging scheme in London is an example of this, where every car entering the center of London is captured automatically on digital camera, its number plate found and read automatically, and compared automatically with a database of cars that have paid the required fee for that day. The situation in the 1960s was very different. Detection of items of interest in a picture was done by a trained photo-interpreter. Given the enormous size of the Soviet Union, a large workforce was involved in reviewing spy satellite pictures.

The wide-area photographs were examined first, and used to detect areas of interest at which to point the high-accuracy cameras next time round.

Nowadays, multi-spectral photographs can be processed on a home PC using freely available software, such as Adobe Photoshop or Microsoft Picture IT. Color contrasts can be changed, dark features lightened, red eyes changed to blue, etc.

Forty years ago this sort of trickery had to be done more laboriously using physical colored filters and the like. In general of course, photographs had to be chemically developed, then printed onto high-quality photographic paper, then examined using microscopes. Digitizing the photographs inevitably produced a loss in detail, so displaying them on a computer screen was not an option.

Despite these difficulties, one of the early pioneering USA satellite photo-interpreters, David Doyle, remembers that "we were seeing areas of the Soviet Union that had not been seen since World War II—we found cities that people had only heard of."[100]

Before the CORONA satellite was available the US knowledge of Soviet missile forces was very sketchy. An official estimate was issued each year by the intelligence community to the Administration and in 1958 this estimate reckoned that the Soviets could have 500 ICBMs by 1962. That figure was based on little hard intelligence and was an estimate derived from analysis of Soviet production capabilities. The 1959 estimate a year later was lowered a bit in the absence of intelligence of actual missiles, forecasting 250 to 350 in 1962.[101]

A force of that many Soviet ICBMs would be able to wipe out the bases of America's strategic bomber forces, and much more besides. These estimates were of course ultra-secret, but the gist of the estimate was released to influential figures in the media, such as columnist Joseph Alsop, and politicians of the appropriate ilk, such as Senator Stuart Symington. Mindful of the apparent Soviet missile lead demonstrated by Sputnik, the public readily accepted the idea that the Soviets had opened up a *missile gap* with the USA. Presidential candidate John F. Kennedy made much of the missile gap in his campaign against his Republican opponent Richard Nixon, which strengthened the "America falling behind the Soviets" argument triggered by Sputnik (Chapter 1) and helped him to win a narrow victory in the 1960 US Presidential election.

One of the last forecasts before CORONA images became available was issued in mid-1960 and illustrated an issue that President Eisenhower had worked hard to avoid, without success. The estimate of Soviet ICBMs differed in the evaluation of the four intelligence bodies involved. The Air Force forecast 700 Soviet ICBMs by 1963, while the Navy and the Army agreed on a forecast of 150, and the CIA predicted an in-between figure of 400. The tendency of the Air Force to take a worst case view of what little information they had was what had impelled Eisenhower to give the U-2 and CORONA programs to the CIA to manage. Eisenhower had not forgotten the non-existent *bomber gap* of the mid-1950s (mentioned above) that had justified the Air Force's expansion of their strategic bomber force.[102]

The low estimates of the Army and Navy were also a consistent feature. Funding to the Air Force to counter Soviet ICBMs would mean less funding for Army and Navy projects—tanks, aircraft carriers, submarines, battlefield helicopters, and the like. Hence Army and Navy estimates tended to take an optimistic view of any Soviet

[100] Day *et al.* (1998) p. 226.
[101] Richelson (2002) p. 27.
[102] Richelson (2002) p. 27.

Figure 28. August 1960: President Eisenhower inspecting the first successful recovery capsule; it contained nothing more than an American flag, but led a week later to the first successful CORONA mission. Others in the photo are (l to r): Colonel Lee Battle, Discoverer Program Manager; AF Chief of Staff General Curtis Le May; General Bernard Schriever, head of the AF R&D Command; Secretary of the AF Dudley Sharp; an unidentified CIA official; and head of the AF Strategic Command General Thomas Power—Le May and Power feature in Chapter 3

capabilities that might require an Air Force counterbalance—while of course adopting a pessimistic stance on issues that would require investment in their programs.

The first successful CORONA mission in August 1960 came too close to the Presidential election to influence its outcome (Figure 28). It took time for the images to be turned into intelligence, especially at first, but once the photo-interpreters had worked out how to read the photos the information they revealed was dynamite.

The first intelligence report incorporating the new satellite data was issued in September 1961. By that time information from three CORONA capsules had become available, in addition to the first whose intelligence value was limited. The new report incorporated new data primarily from CORONA, although the contribution of the recent Soviet defector Colonel Oleg Penkovskiy provided an independent source that added strength to the satellite-derived numbers. The report noted that the Soviets currently had fewer than 25 missiles, and forecast that this would increase to about 100 by 1963—remember the Air Force estimate of 700 a year earlier? Nine months later a new report reduced the number of actual Soviet ICBMs to fewer than 10, of which only 6 were on operational launch pads. The *missile gap* did exist, but it was in America's favor![103] A crash program to deploy hundreds of American missiles to counterbalance the Soviet threat was not needed, saving billions of dollars.

President Lyndon B. Johnson admitted the US Government's debt to CORONA in 1967 when in a famous lapse in security he told a group of officials in Nashville, Tennessee, that "we've spent 35 or 40 billion dollars on the space program and if nothing else had come out of it except the knowledge we've gained from space photography, it would be worth 10 times what the whole program has cost. Because tonight we know how many missiles the enemy had and it turned out our guesses were way off. We were doing things we didn't need to do. We were building things we didn't need to build. We were harboring fears we didn't need to harbor."[104]

The first successful CORONA mission in August 1960 orbited the earth 17 times in a little over a day passing over the Soviet Union 7 times before ejecting its film capsule to be caught in mid-air by the recovery aircraft over the Pacific Ocean off Hawaii—the pilot catching it with his C-119 aircraft on his third pass under the descending capsule. A few days later an audience of photo-interpreters at the CIA was gathered to hear about the latest imagery they would be analysing from over the Soviet Union, not aware that the source was CORONA. They were used to being presented with a map inscribed with a line showing the route of a U-2 aircraft. This time they were presented with a map containing seven vertical stripes and they knew immediately that these represented the portions of the Soviet Union over which a satellite had flown. The audience cheered.

That mission produced over 1,400 photos in which they could identify a number of already known military complexes plus several new objects of interest. The resolution was about 10 m and thus not good enough to provide detailed intelligence. But it produced some new information, including the identification of 20 new SA-2 anti-aircraft batteries (or Surface-to-Air Missiles, hence the "SA" designation and the commonly used SAM acronym) and 6 more under construction, 64 new airfields, and many previously unknown urban areas. SAM sites were a sure sign of something of military value in the area, and thus provided targets for future CORONA missions. This first CORONA also took images of the Kapustin Yar missile test range in the south, 100 km from Volgograd (better known by its former name of Stalingrad), and the Sarova nuclear research complex 350 km east of Moscow. Such images of already known sites were valuable since they could be compared with previous images of the same area and thus alert analysts to changes of potential military significance—new roads, railways, major factory facilities, etc. In turn, they became a reference against which future images could be compared.[105]

Despite the elation at the vast quantity of new material discovered by just one flight, there was also disappointment that the resolution of about 10–15 m was not good enough for hard intelligence information, such as counting the number of bombers.

[103] Day *et al.* (1998) pp. 23–24.
[104] Burrows (1986) p. vii.
[105] Lindgren (2000) p. 102, and Day *et al.* (1998) p. 61.

The second successful flight in December 1960 was of similar quality, but coming as it did in mid-winter with the accompanying short days and widespread cloud cover, the information gained was not as massively novel as for the August mission. The missile and satellite launch facilities at Tyuratam (Baikonur) were imaged as was the Sary Shagan site with its mysterious and massive radar complex—both of these locations being in Kazakhstan.

The next successful CORONA missions were in the summer of 1961 equipped with an improved camera, and they laid to rest any doubts about the missile gap. The first two missions had suggested that the Soviets lacked a sizable force of ICBMs and this was now unambiguously confirmed. Not only was the number of deployed ICBMs fewer than 25, the type of missile was such that it could not be deployed widely requiring cumbersome transportation and support facilities. Thus, the Soviets would not be able to increase the number of ICBMs significantly anytime soon, enabling the forecast of Soviet ICBMs for 1963 to be reduced to 75–125.

These missions identified five ICBM sites under construction of which three had been identified by U-2 flights and signals intelligence but two were previously unsuspected. All five sites were in isolated densely wooded areas and thus difficult to spot from ground level. The SA-2 anti-aircraft missile emplacements being built around them were one of the give-aways from above. ICBM sites are large facilities with rail connections, several hard-to-disguise launch pads, and sizable support buildings and constructions for fueling, erection, and storage of the missiles. Each type of missile had its own specific arrangement of buildings, support equipment, and SAM batteries, and by August 1961 the photo-interpreters had figured out how to distinguish different missile deployments—SS-6, SS-7, and SS-8 were all ICBMs, and SS-4 and SS-5 were so-called medium range (Figure 29).

An issue that would cause difficulties for the SALT negotiations later was the inability to tell whether spare missiles were housed in the support buildings. The imagery could be magnified up to 500 times using microscope viewers revealing the full detail provided by the camera and film. But because of gaps in the coverage since the previous image of the same location, the US could not be sure whether additional missiles were stashed away in the side buildings. During 1961 there was fewer than 20 days of CORONA imaging over the Soviet Union. In 1962 this rose to about 50 days. This coverage left enormous gaps in cover about what the Soviets were doing. And on those days when a CORONA satellite over-flew the Soviet Union it did so for only a few minutes per location, and at a time that the Soviets could predict. Who knew what they might have covered up when a CORONA was due overhead or in the long gaps between satellites? And CORONA was blind if the sky was cloudy or dark, which it was for long periods in the northerly climate of the Soviet Union especially in winter.

These practical limitations made the final count of deployed missiles uncertain by a significant fraction—and the uncertainty could be used by interested parties to justify numbers that supported their particular preferred course of action—the Air Force, for example, might use a pessimistic number to justify an investment in anti-ballistic missile technology, while the Navy might do the same to justify Submarine-Launched Ballistic Missiles (SLBMs), and the State Department might

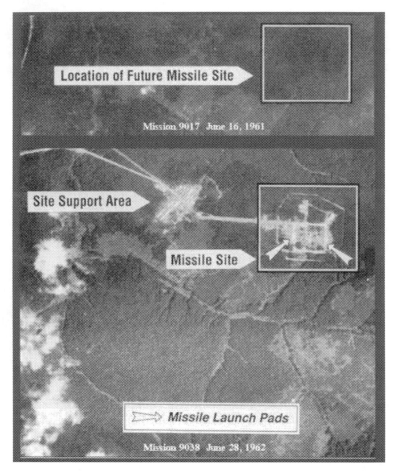

Figure 29. Annotated CORONA images of SS-7 ICBM site at Yur'ya, June 1961 and June 1962. Credit: National Reconnaissance Office.

use an optimistic value to justify a particular negotiated arrangement with the Soviets. The professional engineers and analysts were particularly incensed by the blatant bias in this use of the information they provided. As a former NRO adviser put it "While I agreed that the 'absence of evidence' should not be viewed as 'evidence of absence', at the same time neither is it evidence of the existence of a secret program." He then went on to advise that "there should be healthy skepticism when a participant making an assertion of a worst-case scenario also happens to be interested in funding his own counterpart program."[106]

The CORONA missions in summer 1961 also identified a large number of shorter range missile sites. About 50 sites of Medium Range Ballistic Missiles (MRBMs) with a range of 1,000 to 2,000 km were identified in the western Soviet Union, each with

[106] McDonald (2002) p. 21.

four launch pads on average. Allowing for the proportion of the country that was cloud-covered during the CORONA over-flights, the Soviets were estimated to have 300 MRBM launch pads pointed at Europe. One or two spare missiles were thought to be housed in the storage facilities around each complex. Another 50 or so MRBM sites were identified west of the Urals, and thus targeting China plus US bases in east Asia.

Finally, these 1961 flights identified the early phases of the deployment of an Intermediate Range Ballistic Missile (IRBM) with a range of 3,000 km. The number of roads and support buildings being built led the analysts to estimate that 400 of these IRBMs might be deployed within two years.[107]

The teams of photo-interpreters divided themselves into groups of specialists to deal with the vast quantities of imagery that CORONA supplied. Some would look at the broad picture to spot scenes of interest, another would measure the precise location of items of interest. Others would find supplementary information such as maps to help the interpreters. Specialists in nuclear, naval, missiles, air defense, and other topics would examine scenes of interest to them. They followed the same target from mission to mission, building up a detailed knowledge of how the area was changing.

The mountain of film returned by CORONA compared with the U-2 led to long hours for the photo-interpreters. The government officials and other users of the imagery always wanted more, and more precise, information. International crises would trigger a peak of activity focused on the particular country or region. Robert Kohler recalls that during his years at the CIA one of his teenage daughters was convinced that he was a government assassin. "She was about 15 at the time and could see no reason why the CIA needed engineers. For about two years she was convinced that I was a killer and that the reason I was being called out in the middle of the night was to 'take care of somebody'."[108]

The Soviets deployed decoys to try to fool the satellites. Wooden aircraft, rubber ships, empty silos, and more were used. Most were discovered as time went by. The lack of tracks following snowfall often showed when a site was a decoy. A rubber submarine broke in two following a storm.[109] Former CIA image-analyst Dino Brugioni asserts that "a heavy snowfall negates all camouflaging efforts." He goes on to explain that the pattern of clearing the snow is a give-away, with the most important facilities cleared first, presumably the headquarters, and others in descending order of importance—tracks to latrines are apparently high on the priority list. Melting snow on a roof suggests the building is occupied, while its absence suggests the opposite.[110] Brugioni also says that the best time to over-fly a garrison is Sunday morning when most of the equipment is at home.

Stereo images were a great boon to the photo-interpreter once they became available in the USA from 1962 onwards. They clarified many questions left

---

[107] Lindgren (2000) pp. 106–107.
[108] McDonald (2002) p. 228.
[109] Lindgren (2000) p. 142.
[110] Quoted in Day *et al.* (1998) p. 219.

unresolved by the mono images—for example, turning lines on the image into ditches or fences. The photo-interpreters catalogued railway lines, roads, and power grids. They took note of cables laid between a group of missile silos, indicating the command and control arrangements.

CORONA didn't only count ICBMs and submarines. During the Berlin crisis in late 1961, CORONA images were able to confirm that the Soviets were not massing to invade the west, thus allowing President Kennedy to make a measured response.

TOO LATE

But the time delay in getting CORONA images to the interpreters was a weakness. In August 1968, images of Soviet tanks massing outside Prague were captured by CORONA, but not seen by the analysts until the invasion was over. The images even showed the white crosses the Soviets had painted on their tanks to distinguish them from the identical tanks used by the Czechs. The images also showed Soviet air transport planes assembled near the border, which were later used to fly Soviet troops into Prague. If these images had been received before the invasion the white crosses and the transport planes would immediately have alerted the US to what the Soviets planned.

A year earlier the same thing had happened with the Six-Day War in the Middle East. A combination of a poorly performing satellite and one whose capsule was recovered after the event meant that the US spy satellite images of the Israeli forces preparing to invade Egypt, Jordan, and Syria were not available until it was over.

The US advance warning of the first Chinese atomic bomb was almost the same. Based on CORONA images it was clear that the Chinese test site at Lop Nur was being readied for a test. However, the specialist analysts trying to follow the Chinese manufacture of the bomb were slow to realize how far along the Chinese were with enriching uranium—they were focused on Chinese attempts to use plutonium. The US policy was to announce a Chinese explosion before it happened in order to reduce its propaganda impact. On September 29th after some hesitation the US State Department announced that a Chinese nuclear test "might occur in the near future." On October 15th the CIA changed its official stance and issued an internal memo that predicted a test *within the next six to eight months*. The test took place the next day, October 16th 1964, so in a sense they got it right.

The CORONA satellite images had shown the unmistakable 100 meter high tower on top of which the bomb would be placed for the test (Figure 30). In addition, aircraft movement at the test site had fallen after completion of the facility a year earlier and had picked up a month before the test—clear signs of what was about to happen.[111] This episode showed the difficulty in monitoring the production of uranium, plutonium, or the bomb itself inside buildings such as enrichment plants and nuclear power stations, while demonstrating the power of satellite imagery to detect the preparations for the test.

[111] Richelson (2002) pp. 77–78.

Figure 30. GAMBIT KII-7 image of China's nuclear test site at Lop Nur, 20 days before the December 28th 1966 test. Credit: National Security Archive.

Monitoring a country's production of weapons-grade plutonium or uranium remains as critical a task today as it was in the 1960s. Uranium for civil nuclear power plants is acceptable, but enriching it beyond the point needed for power generation indicates a country's intention to develop nuclear weapons. Plutonium only has one use—making bombs. As we know from current news in Iran and North Korea it is not easy to monitor these materials accurately. Satellites can help by keeping track of the volume of ore being extracted at uranium-mining operations, and then by monitoring the transport of the ore to a processing plant that separates the relatively small proportion of uranium from the usually phosphorus-rich rock. One of the ways satellites can track the uranium is that chemical compounds rich in phosphorus can often be identified in multi-spectral images (images taken in several infra-red wavelengths or "colors"). Even the small amounts of such material that spill from the sides of the railroad trucks along the side of the track can be enough for detection.

The processes inside an ore-processing plant are difficult to monitor from above, but sometimes the plume from the factory chimney can be analysed by multi-spectral imagery to identify what's going on. The slurry left over from the processing of the ore can also be analysed with multi-spectral imagery when it is cast aside as waste.

Enrichment of uranium to make it suitable for a bomb requires huge hi-tech facilities and a lot of electrical power. As mentioned in Chapter 3, the two most favored techniques for increasing the proportion of U^{235} over that of U^{238} are by spinning it in centrifuges or by diffusing it through membranes. Even if a country gets hold of the hi-tech equipment needed for either of these techniques, it then needs a dedicated power station to provide the electricity to drive the equipment. This combination of electrical power and large special purpose factory with military-grade security access restrictions is the sort of combination satellite images can detect or at least highlight for further detailed monitoring.

Plutonium is made in nuclear power stations, obtained by separating it from the spent nuclear fuel. This separation step is a dangerous and highly technical task, requiring specialist facilities and equipment. Satellites can of course identify new facilities, but other forms of intelligence are probably required to clarify what is going on inside. The UN's nuclear watchdog, the International Atomic Energy Agency (IAEA), has powers under the Nuclear Non-Proliferation Treaty to inspect nuclear facilities in order to address this issue and that of uranium enrichment.

The CORONA satellites improved immeasurably the accuracy of counting the number of ICBMs actually deployed. Where it was less successful was in enabling accurate forecasts of future Soviet intentions. In 1961 CORONA put to bed the myth that the Soviets had a lead over the US in ICBMs, demonstrating conclusively that previous US forecasts of Soviet ICBM numbers were much too high—wildly exaggerated might be a more accurate description. From 1965, however, the US would begin to *underestimate* future Soviet deployments.

From 1966 to 1967 the US had new and better spy satellites available—the CIA's wide-area CORONA KH-4B and the Air Force's high-resolution GAMBIT KH-8. The KH-8 in particular provided a new level of detail for the image analysts. It could be maneuvered over a target area of interest (this took days, rather than the minutes Hollywood would have us believe), and as mentioned in Chapter 4 had a combination of black and white images with resolution better than 10 cm, infra-red images of poorer resolution, and multi-spectral images for chemical analysis.

A debate within the intelligence community had established in 1964 that imagery with a resolution of 1.5 m provided no additional intelligence compared with images of 3 m resolution. This conclusion led to cancelation of the development of a CORONA replacement that would have had 1.3 to 1.7 m resolution.[112] Instead, the Air Force spent big bucks and built the 3-ton GAMBIT KH-8 and launched it on the larger (with matching price sticker) Titan-3B rocket.

The photo-interpreters expressed a desire for the wide-area satellites to have 60 cm resolution, but it was to be 5 years before that could be achieved with the Big Bird KH-9. Meanwhile, the CORONA KH-4B was an incremental improvement over its predecessors providing reliable quality with about 2 m resolution, and better motion compensation that allowed the satellite to be maneuvered into various orbits including as low as 130 km. Color film was flown on a couple of missions but the

---

[112] Richelson (2002) p. 123.

worse resolution (twice that of black and white) outweighed the occasional benefits of detecting camouflage, so black and white remained the standard.[113]

THE DARK SIDE

The American and Soviet spy satellites had a generally stabilizing effect on politicians in both countries. Debate about the other's intentions was at least founded on concrete facts about the current state of military deployment, even if forecasting the adversary's future intentions remained a guessing game. One of several sobering ironies in this affair was that the rockets that carried the politically calming CORONA and Zenit satellites into orbit were the very same rockets that in the form of ICBMs were a major cause of the destabilization to begin with.

The spy satellites enabled the other side's ICBM sites to be counted, thus providing reliable information to the policy makers. But another of the ironies was that the same images allowed the ICBM sites to be accurately located and entered into the targeting computers of the other side's ICBMs and long-range bombers. This *dark side* of spy satellite images as a source of target information for a future attack is the reason so much of the information they collected was secret—and remains secret today.

Because of this targeting ability of spy satellites, until the first successful Zenit flight in summer 1962 the Soviets complained bitterly to America about the CORONA flights. Perhaps the Soviets still smarted from the riposte of France's President De Gaulle at the May 1960 Paris summit meeting, 3 months before the first successful CORONA mission. As mentioned in the last chapter, the Summit took place just after the shooting down of an American U-2 plane deep inside the Soviet Union and the capture of its pilot. Soviet leader, Nikita Khrushchev, launched into a tirade directed at President Eisenhower about the US over-flying his country, whereupon De Gaulle interjected that he too had been over-flown. "By your American allies?" asked a surprised Khrushchev. "No, by you," said De Gaulle. "That satellite [Sputnik-4] you launched just before you left Moscow over-flew France 18 times. How do I know you don't have cameras onboard that are taking pictures of France?"[114] This amusing retort had important consequences, because while effectively ending US aeroplane over-flights of the Soviet Union, it gave *de facto* approval to satellite over-flights.

Accurate targeting required two fundamental types of physical information. First, you had to establish the precise geographic coordinates (latitude, longitude, and height) of the launch site and of its target. Then you had to know the variations in the earth's gravity along the trajectory of the missile. Satellites played an essential role in establishing both of these.

[113] Day *et al.* (1998) p. 82.
[114] Ambrose (1984) p. 579.

A map is traditionally created by surveyors measuring the distance and angle between way-points. Widely separated points—with a body of water between them, for example—are tied together using star and sun sightings. On the face of it, if a target was found in a CORONA image then you knew where it was. In practice, you only knew approximately which way the CORONA satellite was pointing when it took the photo. Then you only knew the orbit of the satellite approximately. Next, the time at which the image was taken was only known approximately—remember the satellite travels 8 km a second, so a couple of seconds error in when you think the image was taken gives a 16 km error in location. You also only know approximately how to convert distances on the image into distances on the ground.

If you are like me, you have lots of vacation snapshots taken at an unknown location. By putting the date and time on each photo you could probably narrow down the location quite a bit—if it's Tuesday it must be on the road from Tucson to Las Vegas and the time would put us about half-way between them. It would be better if your camera had a GPS receiver in it so that it could imprint the image with your exact location. If it also included a compass it could indicate the direction the camera is pointing when you took the picture, and that would pretty much tie down what scene you had captured on film (or disk).

Gradual improvements were incorporated into CORONA to address this problem. By the time the first CORONA KH-4B was launched in August 1963, a lot had been achieved. Special cameras were added to detect stars and the horizon to give accurate information about the satellite's pointing direction—the equivalent of the compass in the vacation snapshot analogy. A special very stable radio beacon was added to allow ground terminals to measure the Doppler frequency shift caused by the satellite's movement across the sky—this proved to be the most cost-effective and reliable way to get accurate information on the satellite's actual orbit. A more precise timing system was added to the satellite to give an accurate time stamp for each image frame. Then with the accurate knowledge of the orbit the time stamp told you where the satellite was at the time the shutter was opened—the equivalent of the GPS information in the snapshot analogy (GPS wasn't available until the mid-1990s).

In parallel with improving the CORONA satellites, research groups were busy processing the orbit information from dozens of satellites to work out the precise form of the earth's gravity field. Although the earth is generally spherical and of uniform density, it does depart from a perfect sphere enough to divert a long-distance missile or a satellite by many kilometers. The biggest non-spherical feature is, of course, the equatorial bulge—the earth is 42 km fatter at the equator (from Brazil to Borneo, say) than it is at the poles due to the centrifugal force of its rotation. Smaller gravity features vaguely follow the contours of the continental plates and the mid-ocean ridges, but the detailed picture is highly complex. By looking at the small deviations in thousands of orbits of hundreds of satellites over several years, researchers calculated the details of the gravity field. Armed with this mathematical model of the earth's gravity, the future trajectory of a missile or rocket can be accurately predicted.

Research groups in the US Army, Navy, and Air Force worked on this problem of precisely characterizing the earth's gravity field and shape throughout the 1960s.

So, too, for purely scientific purposes did several groups funded by NASA, and I had the good fortune to be part of one of those when I first came to the USA in 1966. That group, led by NASA's John Berbert, compared the trajectories calculated by all of these civil and military communities of a special research satellite called GEOS (or Explorer 29) that NASA had launched in November 1965 for the purpose of improving the ability to compute a satellite's orbit. Each of the research groups sent their trajectory calculations to NASA and those from the Army had one peculiar feature. While everyone else measured azimuth starting with due north as zero, the data from the Army group at the Redstone Arsenal in Huntsville, Alabama, used due west as the zero azimuth point. We jokingly theorized that this was due to the influence of the Huntsville boss, Werner von Braun, for whom due west had been the direction to fire his V2 rockets to hit Britain from his war-time base at Peene-münde in Germany.

You still have to convert a distance in centimeters on the image into kilometers on the map, and the panoramic form of the CORONA lens complicated this task— like trying to work out the actual location of objects in a fish eye photo. To figure out what exactly each CORONA camera was "seeing", target areas of the US were photographed by each satellite and the images compared with accurate maps. Then when the satellite photographed another part of the world the image could be converted to a map by comparison with the US target areas—a process called "calibration" of the cameras.

By the end of the 1960s maps of all parts of the world could be generated from CORONA and GAMBIT imagery accurate to about 100 m in the worst case—and, of course, much better in areas that had been studied in the most detail. These sorts of accuracies were adequate for targeting nuclear missiles whose kill zones were many times greater than the mapping errors.

Since the main part of this story concerns the 1960s and 1970s, we will delay until Chapter 10 how spy satellites provide another form of targeting information specially designed for cruise missiles—a delivery mechanism that didn't come into widespread use until the 1980s.

6

The road to SALT-I

By the late 1960s leaders in the Soviet Union and America had come to understand that enhancing either your defence against ICBMs or the ICBMs themselves led to a *reduction* in your security. Enhancing your offensive weapons motivated the other side to enhance theirs. And enhancing your defences motivated them to enhance their offensive weapons, thus worsening the problem. So, improving your defences decreased your security. That is the peculiar logic of détente. The stage was set for negotiations between the two superpowers to find a way out of this catch-22.

They also realized that it was one thing to agree not to deploy a new set of weapons but much more difficult to roll back weapons that were already in the field. It was politically and administratively more difficult to write off an expenditure that had already been made or committed than not starting a new expenditure.

Another thorny issue was how to verify that the agreement was being respected (i.e., that the other side wasn't cheating). The Soviet dictatorship thrived on secrecy, especially concerning its own citizens, and this was not a sound basis for open and frank negotiations with an adversary. Trust between the two countries was in short supply. As we will see below, Henry Kissinger seems to have got some enjoyment out of telling the Soviet diplomats things about their own missiles that their military hadn't told them.

In this chapter we take a look at how the immense nuclear over-kill of both the US and the Soviets gradually overcame their mutual suspicion and brought them to the negotiating table. First, we look at the agreements on other things such as polluting the atmosphere with radioactivity and (having outlined the design of a hydrogen bomb) on stopping small countries getting hold of nuclear weapons. We also analyse why there is an arms race at all, given that no one seems to benefit from it

THE BAN ON ATMOSPHERIC TESTING

The events surrounding the ban on atmospheric testing of nuclear weapons illustrate how difficult it was to break through the mutual suspicion. As outlined in Chapter 5, the US and the Soviets issued separate statements in 1958 declaring that they would not be the first to test a nuclear bomb in the atmosphere. In 1961 the Soviets re-started atmospheric testing claiming that France's tests the previous year had invalidated the 1958 declaration. The US soon resumed atmospheric testing itself, citing the Soviet resumption as the cause.

The objective of these tests was to develop a light-weight bomb for delivery by missile or plane. The first US hydrogen bomb, codenamed *Mike*, used super-cold deuterium involving large cooling pumps and Dewar vacuum flasks to maintain the low temperature. It was bulky, fragile, and heavy—65 tons. Its explosion on Eniwetok Atoll on November 1st 1952 (October 31st in the USA) was the first ever man-made explosion with a power in excess of a million tons of TNT—a megaton. Forecasts of the power the bomb would achieve were very uncertain—between 1 and 10 megatons, with an outside chance of it reaching 50 megatons. In the event its yield was 10.4 megatons—the ground-level fireball extended 5 km across compared with 300 m for the Hiroshima atomic bomb and the mushroom cloud topped out at 50 km high. And it was a very dirty bomb—three-quarters of the power, 8 megatons, was from fission of uranium and plutonium (in other words an atomic bomb), and only a quarter came from the fusion of hydrogen into helium. Its peculiar design involved the X-rays from a plutonium atomic bomb (Nagasaki-style) compressing a layer of uranium which in turn compresses the deuterium which in turn compresses an inner-most layer of plutonium (called the sparkplug) which reaches critical mass. The atomic bombs inside and outside the deuterium (which had been seeded with a few grams of tritium, the rarest of the isotopes of hydrogen) caused the fusion of the deuterium. Nobody knew for sure if you could make hydrogen (in its deuterium variant) fuse to become helium in a sustained way—the theory said it should happen and the sun and the stars were powered by this process—but no one had ever done it. So, Mike's designers took no chances and used the fission–fusion–fission scheme developed jointly by Stanislaw Ulam and Edward Teller (for which the erratic, brooding, and arrogant Teller is usually and erroneously given the primary attribution, even though his contribution seems to have been at best secondary) to ensure that the deuterium ignited (Figure 31).

The first Soviet hydrogen bomb (H-bomb) test came just 10 months after Mike on August 12th 1953. Like Mike most of its power came from its uranium and plutonium fission actions, and only 15–20% from the H-bomb principle of fusion. Its yield of 400 kilotons was much smaller than Mike's, primarily because the Soviets hadn't figured out the Teller–Ulam principle of using atomic bomb radiation to compress the deuterium—they used conventional high explosives to do that instead. But they were ahead of the Americans in using a room temperature form of deuterium rather than the bulky and fragile ultra-cool form. The trick was to use lithium deuteride (the deuterium variant of lithium hydride) which is a solid at room temperature, whereas deuterium (which chemically is identical to hydrogen) is a

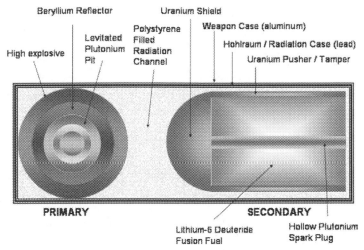

Figure 31. Schematic of the Teller–Ulam hydrogen bomb; radiation from the primary atom bomb (left) ignites the second atom bomb or spark plug at the center of the secondary (right) so that the lithium deuteride (right) is compressed from both outside and inside ensuring that it fuses; the first US H-bomb used liquid deuterium in place of lithium hydride.

gas—the ultra-cooling of deuterium in the Mike bomb was to liquefy it and thus make it sufficiently compressed to ignite. The Soviet design was derived from information provided by Klaus Fuchs before he left the US nuclear center in Los Alamos in 1946 (Figure 32). He had passed over information on the so-called "alarm clock" H-bomb design favored by Edward Teller, and to which Fuchs had made important contributions (we will come back to this below). Despite doubts by several colleagues, especially Stanislaw Ulam, that the design would work, Teller insisted it was the best approach. The Soviets therefore started from where the US was in 1945–1946, whereas by 1950 the US had moved on to a newer and more flexible design—the so-called staged radiation implosion design used in Mike.

The US was pursuing a lithium deuteride bomb too, but had chosen to build the Mike design first in order to be sure to get a fusion reaction. News of the Soviet H-bomb test so soon after the first American test released a wave of anti-communist feeling across the US, driven by the realization that Fuchs and other spies had passed so many valuable American nuclear secrets to the Soviets. Los Alamos hero, Robert Oppenheimer was investigated and pilloried, and Senator Joe McCarthy's Un-American Activities hearings gained strength; McCarthy issued a statement the following April that communists in the US government had delayed US research on the H-bomb by 18 months—as always, McCarthy lent an air of authority to his assertions by including a precise but spurious numerical value. Seven months after the Soviet Joe-4 H-bomb (Joes 1 to 3 were atomic bombs) the US Bravo bomb was exploded on Bikini Atoll on March 2nd 1954. Once again, the designers loaded the dice every way they could think of to ensure that fusion of deuterium into helium actually occurred. They enriched the lithium, in the same way uranium has to be

Figure 32. Photo of Klaus Fuchs from his Los Alamos security clearance.

enriched for a uranium atomic bomb, so that 40% of it was lithium 6 (Li^6) which fuses much more readily than the more common Li^7.

Although it weighed only $10\frac{1}{2}$ tons, and was thus capable of being delivered by an aircraft, Bravo showed that the theory had still not caught up with practice. It was expected to be a 5-megaton device but actually yielded 15 megatons. Some time later the scientists realised that they had overlooked an important source of fusion energy from Li^7, which explained their Bravo under-estimate. This was the largest ever US nuclear explosion, with a fireball that measured 6.4 km across (remember Hiroshima's 300 m). Bravo's huge size resulted in radioactive fallout on a far wider scale than expected. The 23-man crew of the Japanese fishing boat *Fukuryu Maru* (Lucky Dragon) were badly contaminated while fishing 130 km away, and one of the crew later died from his injuries. The US offered help with treating the radiation sickness but would not release details of the type of radiation in case the Soviets learned that Bravo was a lithium deuteride device. Coming after the nuclear incineration of Hiroshima and Nagasaki, this third American nuclear attack on Japan (as the anti-American media painted it) started the public outcry that resulted nine years later in the ban on atmospheric testing.

The Soviets paid some attention to minimizing radioactive fallout on civilians, but the policy was spotty at best. Soviet H-bomb guru and later peace campaigner Andrei Sakharov recalls that the local Kazakhstan population was evacuated before the first H-bomb test, Joe-4. But near the nuclear facilities at Chelyabinsk, the reactors discharged their highly radioactive waste products directly into the Techa river. The laborers were not warned about the dangerous levels of radiation in the rivers and lakes (apart from a few "swimming forbidden" notices), and many received radiation doses high enough to cause severe illness. The workers inside the plants also received dosages way above the safe limit, especially during the construction phase. Many were political prisoners who had their sentence commuted by two years for every year worked in the difficult and dangerous conditions—but they weren't allowed to leave when their sentence was over because of the risk of security lapses.[115] Chelyabinsk is said to be "the most contaminated spot on the planet",

[115] Rhodes (1995) pp. 350–351, 514–515.

with the years of casual waste disposal culminating in an explosion in a nuclear waste tank in 1957 resulting in a huge release of radioactive particles. Local people were evacuated but forbidden to speak about the event. The three provinces surrounding the facility were contaminated, forming what is called the East Ural Radioactive Trace.[116]

The third US H-bomb test (Romeo) shortly after Bravo also yielded three times its predicted power for the same reason as Bravo. In due course the US got the theory and the practice to match, and produced relatively light H-bombs (hence deliverable by missile and bomber) with yields in the sub-megaton range that was considered adequate for most military targets.[117] But it took time and lots more tests to get the details right—by the time of the test ban moratorium in 1959, the US had exploded more than 150 atmospheric tests (more than 50 in 1958, alone) and more than 30 underground tests, and the Soviet tests numbered more than 100, almost all of which were in the atmosphere.[118]

The public alarm at the radioactivity created by atmospheric tests forced the leaders of both superpowers to seek ways to halt them. The public groundswell was for *all* nuclear tests to be banned—not just those in the atmosphere. But both the US and the Soviets were not confident that they could detect all underground tests—and their military were insistent that further tests were needed. Both feared that the other side would figure out how to perform underground tests in such a way that they were indistinguishable from earthquakes.

The Cuban crisis had almost resulted in a nuclear war, and gave a powerful impetus to the US and the Soviets to make some progress on controlling the situation. Agreement was reached therefore in 1963 to ban all nuclear tests—except underground. Tests in the atmosphere, underwater, and in outer space were fairly easy to detect and so were forbidden. In an executive declaration the previous year, President Kennedy had endorsed the need to ban atmospheric tests and linked that policy with the need for techniques of verification. He went on to recommend that future treaties be written in such a way that their provisions could be verified—there had to be externally observable features detectable by "national technical means" such as satellites.[119] If you couldn't verify it, you excluded it from the agreement, and this approach was to prove workable for the next 20 years.

Another feature of the test ban treaty that was to reoccur was the difference of opinion within the various parties inside each country. The military argued that whether or not you could verify underground testing, as long as a stock of nuclear weapons was in the armory, it would be necessary from time to time to explode one underground to check that they still functioned. The politicians argued that progress in controlling the spread of nuclear weapons would be difficult if the superpowers exploded them every now and then, even underground. Both of these viewpoints were

---

[116] DeGroot (2005) p. 199.

[117] Details of Mike, Bravo, and Joe-4 are from Rhodes (1995) pp. 482–512, 523–525, 541–542.

[118] Myrdal (1974) pp. 32–33.

[119] National Security Action Memorandum (NSAM) 1156, dated May 26th 1962, quoted in Temple (2005) pp. 353–354.

reasonable and strongly felt by their advocates, so finding a way forward was never going to be easy.

THE NON-PROLIFERATION TREATY

While there were many reasons to expect arms negotiations between the Soviets and the US to fail, such as the Vietnam War and the continuing tensions in Europe, the Middle East, and Africa, there were also positive indications. One of the most important was the Nuclear Non-Proliferation Treaty (NPT), signed in 1968. The cynics saw this Treaty as the major powers wanting to avoid other countries becoming a threat to them. It did indeed divide the countries of the world into two categories—the five nuclear powers, the Soviet Union, the USA, Britain, France, and China, and the rest. It was clear why the five nuclear powers favored that division of the world, and the beauty of the Non-Proliferation Treaty was that it provided an incentive for the other countries to accept their lot in the non-nuclear club. Note that although France and China didn't immediately sign the NPT they agreed to abide by its provisions.

In return for being the only nuclear powers, the big five had to agree to help the other countries in various ways. Economic and technical assistance was promised to developing countries in the use of nuclear energy for peaceful purposes. In addition, the big five promised to pursue negotiations towards nuclear disarmament, and indeed towards complete disarmament under international control. In the 21st century, these two commitments to disarmament (nuclear and general) have become a major source of tension in maintaining the success of the Treaty, and I will return to this in Chapter 10.

Nevertheless, the Non-Proliferation Treaty reinforced the view in all countries that the nuclear arms race could not continue escalating indefinitely. The various nuclear tests in the atmosphere had increased radiation levels world-wide and brought home to everyone the danger of a nuclear exchange involving hundreds or thousands of such weapons.

THE SHIFTING BALANCE OF POWER AND THE CUBAN CRISIS

Most of the military too realized the senselessness of building more and more nuclear weapons and delivery mechanisms. With a few exceptions, they came to appreciate what Henry Kissinger called "the ultimate nuclear dilemma", namely that the strategic arsenals were useful primarily to deter nuclear attacks and for little else.[120]

The long-term shifts in the strategic balance also favored an agreement at this time (the late 1960s) more than any other time since World War II. A sustainable agreement is more likely to be reached when the parties are of roughly equal stature. Since the War, the Soviet Union had been perceived in the West to be superior in

---

[120] Kissinger (1999) p. 119.

conventional arms. This assessment was based ironically on a Soviet Army that was only ever used in anger against its own allies—East Berlin 1953, Hungary 1956, Czechoslovakia 1968, and China (border exchanges) 1969. To balance that, the US was perceived by the Soviets to have superiority in long-range strike forces— bombers and missiles.

Henry Kissinger points out that the Soviet–USA strategic relationship went through four stages between World War II and the signing of the SALT-I agreement. Immediately after World War II America had a nuclear monopoly, and had shown a willingness to use these powerful weapons against an enemy (in Japan). Then in the second stage, from 1950 the Soviets started to build up a nuclear arsenal, but until the 1960s the US had a distinct edge—enough to enjoy what was called a "first strike" advantage. First strike meant that the US could destroy all of the Soviet nuclear forces while incurring relatively low casualties itself. Fortunately, this assessment of their relative strengths was never tested in practice.

The Cuban crisis (see Chapter 3) marked a turning point. Most Western commentators hailed it as an American victory, but historians today are tending to take a more nuanced view. Yes, nuclear weapons were removed from Cuba where they would have been a source of severe tension. But in return the Americans removed their nuclear missiles from Turkey. This American withdrawal was recognized at the time by the US government as something of a climb-down on their part, and they persuaded the Soviets to keep that part of the agreement secret. The US rationalized its decision by claiming that it had been planning to withdraw the missiles from Turkey anyway on the grounds of their obsolescence. There was some truth in that, but the fact remains that the Jupiter missiles in Turkey might well have stayed there another 10 years or more were it not for the *quid pro quo* with the Cuban missiles. And if it was planned all along, why the secrecy?

So, with the benefit of hindsight the outcome of the Cuban affair seems to have been more like a tie than an American victory. However, at the time the consequences were as if it *had* been a US victory. Perceptions and expectations count for quite a lot in politics, and although the Soviets achieved a coup of sorts in forcing the US missiles out of Turkey, that measured view was not how the Soviet leaders saw it at the time. In fact, the apparent USA victory in the Cuban crisis strengthened the forces in the Soviet Union that were pressing to increase all categories of Soviet power.[121]

The resolve of the Soviets to achieve nuclear parity with the USA was hardened by American boasting. The fact that the missile gap actually favored America, rather than the Soviets as had been believed before the first CORONA flights (see Chapter 5), was leaked to the US media, and mentioned in contacts with Soviet diplomats. Those remarks helped to trigger the Soviet adventure in Cuba as a way to close the gap. After Cuba, the Soviets became determined to avoid any similar humiliation in the future, and began their rapid and expensive missile build-up.

So, the third stage of the Soviet–USA relationship began after 1962 with the initiation by the Soviets of a massive program of constructing and deploying

[121] Kissinger (1979) p. 197.

long-range missiles. They replaced their small number of unprotected missiles with 1,400 missiles in concrete silos in fewer than 10 years. They also built up a fleet of nuclear-powered submarines equipped with nuclear missiles. By the end of the 1960s it was clear to most observers that the USA had lost its first-strike advantage—now nuclear war implied unacceptable losses to both sides irrespective of either side's numerical advantage, or who fired first.[122]

We have already seen in Chapter 3 how the steady growth in Soviet strategic power caused a reappraisal in the USA of the policy of mutually assured destruction (MAD), and how it also caused the right-wing activists (the "hawks"—as opposed to the "doves") to argue for a pre-emptive strike by the USA before the Soviets became too strong.

By the late 1960s the Western powers had lost nuclear superiority over the Soviet Union and the fourth stage of the post-war relationship had arrived—rough nuclear parity between the superpowers. Henry Kissinger points out that the US threat to launch a nuclear attack had therefore to some extent become a hollow threat. He also notes that the West failed to compensate for its loss of nuclear ascendancy by building up conventional forces. Thus, overall the Soviets seemed to have gained the upper hand—neutralized the US nuclear lead and retained a lead in non-nuclear land forces. Kissinger envisaged the US becoming subject to nuclear blackmail unless the West built up conventional forces at least locally to resist Soviet pressure.[123]

However, the Soviets were still mindful of America's much superior production capacity. They remembered the US ability in World War II to simultaneously fight a two-ocean war, supply the Soviets with vast quantities of Lend Lease and fund the $2 billion development of the atomic bomb. The quantities of Lend Lease material arriving in the Soviet Union during the War were staggering, and left a lasting impression on the Soviets—the total of $11 billion included 400,000 trucks and 600 naval vessels. The great Soviet drive westwards into Poland and Germany in 1944 and 1945 was on American axles.[124]

THE POLITICS OF ESCALATION AND NUCLEAR PARITY

The cost-effectiveness of that $2 billion (perhaps $25 billion in today's money) spent on the atomic bomb turned out to be dubious. The official US government assessment of the Hiroshima and Nagasaki bombs equated them to 2,100 tons and 1,200 tons, respectively, of conventional bombs—a far cry from the figure of 15,000–20,000 tons of TNT usually quoted. I often wondered why Hiroshima and Nagasaki had been selected for these bombs and not Tokyo or Osaka. It turns out that all the other important targets had already been obliterated by conventional bombing raids, and the two atomic cities were held back from receiving the same treatment deliberately in order that a target would still exist for the new bombs.

[122] Kissinger (1999) p. 116.
[123] Kissinger (1982) p. 259.
[124] Rhodes (1999) p. 95.

Figure 33. At the Big Three conference of February 4-11 1945 at Yalta in the Crimea, Stalin (right) agreed to enter the Pacific war; Churchill (left) and Roosevelt (who died soon after) made important concessions to Stalin on Poland.

But Hiroshima and Nagasaki caused Japan to surrender and thus avoid a million US casualties during an armed invasion, didn't they? It seems not—at least not the way we have been led to believe. The killer stroke was that the Soviet Union declared war on Japan two days before Nagasaki. Japan had no illusions about the ambition of the Soviets to grab large chunks of what Japan considered to be the homeland, and rapidly chose to surrender to America before Soviet advances made this impossible. In fact, the Soviets and Japanese continued fighting for another three weeks (until September 9th 1945) enabling the Soviets to lay unambiguous claim to Manchuria and North Korea—unambiguous that is with respect to the US who Stalin feared would try to grab some of the coastal areas. During these few weeks the Japanese forces suffered 80,000 deaths due to the overwhelmingly better equipment and superior numbers of the Soviet forces. At the Yalta Conference involving Stalin, Roosevelt, and Churchill in February 1945 (Figure 33), Stalin had agreed to declare war on Japan after Germany's defeat in return for Soviet rights to various parts of east Asia (southern Sakhalin, for example). At the end of May the Soviets gave

August 8th as the date when their forces would be in position, and confirmed this at the Potsdam Conference in the second half of July. It is not clear therefore which is chicken and which is egg: the US wanted to drop the atom bomb on Japan before the Soviets entered the Pacific war so as to stake a claim to having been the cause of Japan's surrender. The Soviets wanted to be able to claim the same thing too. What is clear is that the Soviets had no hesitation about entering the war on August 8th once the Hiroshima bomb had been dropped since Stalin was sure that Japan would surrender within a few days in the wake of this city-leveling bomb—Stalin knew all about the US bomb from Fuchs and other spies and was only waiting to see if the Americans would actually use this new weapon. So, in a sense the atomic bombs did trigger the Japanese surrender but only because of the Soviet perception that those bombs would trigger a Japanese surrender.[125]

Nuclear bombs then are weapons of terror, and of powerful political influence out of proportion to their kilo-tonnage, but otherwise of little military value—explaining why they haven't been used in any of the many wars fought by the nuclear powers since 1945—Vietnam, Korea, Algeria (France), Iraq, the Falklands (UK), the Middle East (Israel), and so on.

The $2 billion spent by the US in developing the atomic bombs (most of which went into building the nuclear reactors and uranium enrichment and processing facilities at Oak Ridge and Hanford rather than manufacturing the bombs *per se*) could have provided America with a third more tanks, or five times more artillery or two-thirds more B-29 bombers.[126] Would the war have been shortened more and would fewer American servicemen have died if the money had been spent in one of these other ways? It's certainly possible this would have been the case.

Since nuclear weapons were of no use in any of the wars fought since World War II why did the Soviets and the USA want to keep them? This simplistic question is really quite a difficult one to answer fully, but it strikes at the heart of the matter as concerns the nuclear arms race.

Many of the leaders on both sides claim that they would never have used nuclear weapons. Khrushchev said that he suffered sleepless nights until he came to the realization that he could never possibly use these weapons. Eisenhower said that war with nuclear weapons was unthinkable. Even Truman, who *had* used the atomic bomb against Japan, said in 1953 that such a war is not a possible policy for rational men.

The USA certainly was deterred, and the Soviets probably were too, by the thought of even one of its cities being obliterated by a nuclear weapon. So, why then the 10,000 plus nuclear warheads in the 1970s' American and Soviet arsenals?

The political system in the US seems to be at the root of it. The military services vied with each other for more of the defence "cake" and to be in command of the military "high ground", and missiles and nuclear warheads were very much on the high ground at least politically. Scientists and industry lobbied for new technologies which defence programs could deliver. And Administrations of both persuasions

---

[125] Werth (1965) pp. 873–874, 919, 923–930.
[126] Edgerton (2006) p. 16.

responded to pressure in election years to be strong on defense and to strengthen the alleged pump-priming effect of defense spending. A politician perceived to be weak in the face of the Soviet threat had difficulty getting elected, so it paid (politically) to be a hawk. Let's not forget that there are elections every two years in the American system, so the pressure was pretty much continuous.

The Soviets experienced similar pressures, although in the early post-war years the opinion of the dictator, Stalin, was perhaps the decisive factor. It seems hard to credit it, but in the depths of World War II when the Soviet Union was barely surviving the Nazi attack, Stalin authorized the start of an atomic bomb program. At this point of the war, 1942, Leningrad was in the middle of its horrific two-year-long siege and the battle of Stalingrad was raging. But information from Soviet spies in the USA alerted Stalin to the start of the Manhattan Project to develop the atomic bomb. Despite America being his ally, despite the perilous state of the Soviet economy, and despite Soviet spymaster Beria suspecting that it was all disinformation fed to the Soviets to trick them into useless research, Igor Kurchatov was tasked with duplicating the American work. In the event, wartime shortages prevented the Soviet bomb being ready until 1949, but the sequence of events illustrates the interaction between decisions in the two countries—the politics of escalation where one side develops a new weapon for fear that otherwise the adversary will gain a great advantage through its sole possession.

Much the same thing happened with the hydrogen bomb. During World War II, the scientists in Los Alamos developing the atomic bomb talked about building a "super bomb" based on fusion of hydrogen. They performed some studies and produced some outline designs. However, they were far from having the authorization or resources to begin development—they did enough work to establish that the task was extremely complex and would require an effort similar in scale to that required to develop the atomic bomb. The chief Soviet spy in Los Alamos Klaus Fuchs dutifully passed on information about the super bomb to his Soviet contacts, who in turn felt compelled to pursue research into the fusion bomb. America in fact slowed down work on the H-bomb after the war, judging it too expensive and militarily unjustified, not to mention technically unproven especially in the form advocated by Edward Teller. Then when the Soviets exploded their first atomic bomb in August 1949 three to five years earlier than America had expected (the National Intelligence Estimate seven weeks earlier said the Soviet bomb was four years off), the Americans realized that spies had given the Soviets huge chunks of priceless information. Five months later Fuchs confessed to being a Soviet spy and the US realized just how much information the Soviets had been fed—Fuchs was the holder of a joint patent with the mathematician and computer pioneer John von Neumann of a design for a key part of a H-bomb, and the full bomb design was attached to the patent application. So, the Soviets knew practically everything Los Alamos did in 1945 about how to build a H-bomb. This sparked a full-scale development of the H-bomb in the USA, which in turn justified the Soviets doing the same thing.

The delay in starting the development of the H-bomb in the USA was caused by the lack of clear military targets for such a weapon, or of a means to deliver it to any plausible target (it looked like it would weigh 100 tons or more). But these practical issues were swept aside in a wave of reaction to the Soviet atomic bomb blast and the exposure of how much information Fuchs had given to the Soviets. Ironically, Fuchs also passed on much of the same information to the British when he returned to Britain in the summer of 1946, so his penchant for proliferating nuclear bomb information was not obsessively communist-oriented. Incidentally, since the Soviet Union was not an enemy at the time, Fuchs was not convicted of treason and got the maximum sentence of 14 years, not death as had been widely expected.

After the Soviet's Joe 1 explosion (as the US labeled the event—its Soviet code name was *First Lightning*) in August 1949, the US Air Force suddenly discovered that they needed a hydrogen bomb urgently—as did the US Navy and the Army. If one was going to have it, the others were too. Thus was born the "triad" of the American nuclear deterrent—nuclear weapons in submarines, in land-based missiles, and in aircraft. No doubt if there had been a fourth or a fifth military service, the deterrent would have been a quad or a quin.

In his parting speech on retirement, President Eisenhower warned about the perils of the military–industrial complex, and he was referring of course to the sort of events surrounding the development of the H-bomb. At the time, his warning was written off as the quixotic remarks of an aging and none too bright statesman. Historians have pretty drastically revised their opinion of Eisenhower since then and he is now generally rated as one of the most effective of all American Presidents—presiding over a period of relative peace and unparalleled prosperity. His remarks about the inexorable pressure from the military community are also now more fully appreciated.

THE ABM TREATY

Introduction of a new weapon system by one of the services was always subject to lengthy debate in the US not only because of the high price of such systems but also because of the positioning of the other two services and of the Senators and Congressmen from states wishing to be involved in the development work. In contrast, there was hardly ever a push for eradication of an existing weapon system. So when it came to negotiating an agreement with the Soviets it was always going to be easier to agree *not* to introduce some new weapon than to eliminate an existing one.

This general principle dictated the debate about halting the growth in the number of missiles. Both countries had made major investments in ICBMs, submarines, and airborne missiles, thus any attempt to roll back the results of those recent and massive expenditures was going to be difficult. Deployment of anti-ballistic missile (ABM) systems, on the other hand, had only just begun and was proving technically difficult and very expensive, so ABM systems offered the simplest area on which to agree a halt.

A Soviet ABM system was deployed to protect Moscow. First installed in 1964 it had gone through various make-overs, becoming operational in 1972 a few months before SALT-I was signed. At the same time, the USA was developing an ABM system in North Dakota to protect the ICBM launch sites in that region. Enormous additional expenditure would have been necessary to extend the protection of either of these ABM systems to give full cover of either country. Thus, a decision to halt further ABM deployments was relatively easy compared with halting the production lines of missiles, planes, and submarines.

The early forms of the ABM system had involved a defensive missile that exploded a nuclear bomb in the path of the incoming attacking missile. The nuclear-tipped defensive missile had to detonate its nuclear bomb high enough above the ground that the radioactive fallout never reached the earth. This form of defence meant that the defensive missile didn't have to get particularly close to the attacking missile because the effects of the nuclear bomb were so widespread. The down-side to this approach was that the nuclear explosion caused havoc in the ionosphere (above the atmosphere) making it very difficult for ground-based radars to see anything and interfering with long-distance radio communications. So, any attacking missiles following the first one would arrive with a fair chance of not being detected by your ABM radar. From the defender's point of view nuclear explosions above the atmosphere were bad news—like unplugging all your radars and long-distance radio sets.

So, in the early 1960s both superpowers agreed not to explode nuclear weapons in space as part of the treaty banning nuclear bomb tests in the atmosphere already described at the start of this chapter. That meant that to halt an attacking missile you had to get close to it and either hit it head-on or explode a conventional bomb close enough to it to destroy or disable it. With a nuclear-tipped ABM you only had to be within a mile or two of the attacking missile, but now you had to get within a few meters. The difficulty of course was that the incoming missile was traveling at several thousand km/h, and your defensive ABM missile was doing the same in the other direction. So, the problem became one of stopping a speeding bullet by hitting it with another bullet.

Oh, by the way, the incoming bullet is concealed within a flurry of phoney harmless bullets that are traveling at the same speed and on the same trajectory, and look similar. So, to be on the safe side you should launch a defensive missile against every one of the incoming missiles—phoney and real. The use of a nuclear-tipped interceptor missile would have eliminated this problem since the nuclear explosion would probably destroy all of the incoming missiles, phoney and real, but we've just said that was ruled out.

Both superpowers undertook expensive research programs to find a solution to this problem. Both felt that they had the technology to "hit a bullet with a bullet"— that is, to figure out the trajectory of the attacking missile using radars early enough to launch a very fast missile and steer it to hit the attacker. The failure of the US Patriot systems to prevent Iraqi SCUD missiles hitting Israel and Saudi Arabia during the 1991 Kuwait conflict casts some doubt on this claim.

Even if you believe that an incoming missile can be hit by a defending one, the problem of phoney missiles or decoys remains. An incoming missile could carry dozens, even hundreds of dummy warheads that would be difficult to distinguish from the small number of real warheads. The dummy warheads only had to reflect radar signals in much the same way as the real ones to fool the defender, and this was cheaply and easily done using, for example, strips of metal. In the vacuum of outer space a heavy object moved just like a light one—you may remember the experiment at school of dropping a feather and a steel ball inside an evacuated glass flask and watching them fall at the same rate, or perhaps you have seen the video of Apollo 15 astronaut David Scott performing the same experiment on the moon during his walk near Hadley Rille on August 2nd 1971—a 1.32 kg hammer versus a 0.03 kg feather, hitting the lunar surface together after a 1.2-second fall.

The real missile becomes apparent as soon as it and its cohort of decoys begin to re-enter the atmosphere as they descend towards the target. The decoys burn up while the real missile is protected by a re-entry shield against the high temperatures. So, if you wait long enough the wheat sorts itself from the chaff. But then you have almost no time to intercept the attacking missile and if you somehow manage to do so it will be close to the target which will be subjected to falling debris from the destroyed missile.

In Chapter 10 we will come back to the technical solutions that are just about feasible and affordable now, but in the 1960s and 1970s the required technology was not close to being available.

The first discussions about limiting ABM systems was made by USA Defense Secretary Robert S. McNamara at the USA–Soviet summit meeting at Glassboro, June 23–25 1967,[127] but Soviet Premier Kosygin rejected the idea on the grounds that ABMs were defensive. Indeed, he told President Lyndon Johnson that "giving up defensive weapons was the most absurd proposition he had ever heard."[128]

The Soviets had a point in that their deployment around Moscow was much less provocative and destabilizing than the proposed American ABM in Grand Forks, North Dakota. The Russian ABM would protect human lives but would not prevent Soviet ICBMs being destroyed. The American deployment, on the other hand, was intended to protect American ICBMs so that they would survive a nuclear attack, and allow America to strike back—it was the equivalent of deploying many more missiles and just as destabilizing.

The original rationale for the USA ABM was to defend against attack by third countries (China, for example) and perhaps against accidental Soviet launches. It was therefore supposed to protect against just one or two missiles not the hundreds or thousands the Soviets would launch in an all-out first strike attack. There was wide agreement that the obvious place to site such an ABM system was either around Washington DC or in the northwest of the continental USA where an incoming missile from east Asia (i.e., China) would first appear. However, the US Administration judged that a proposal to protect Washington DC would be seen by voters in the rest of the country as blatant self-interest—politicians protecting themselves and leaving tax payers undefended. Likewise a proposal to site an ABM system in, say, Oregon or Washington states would fail to get Congressional approval because every other state would demand similar safeguards.

[127] York (1983).
[128] Kissinger (1979) p. 208.

Figure 34. The Anti-Ballistic Missile (ABM) complex at Grand Forks, North Dakota.

In the end, the arrangement that *did* get Congressional approval was to site the required giant ABM radar near Grand Forks, North Dakota, so as to defend ICBM sites in that general area (Figure 34). Politically, this was acceptable since if North Dakota was willing to host ICBM sites and thus become a target for Soviet missiles, it deserved an ABM system to protect it. Once the Grand Forks radar and the associated interceptor missiles had been demonstrated to work, politicians in other states that hosted ICBM sites could expect to receive the same protection.

Despite the initial Soviet rebuttal of an ABM treaty, American spokesmen persisted in arguing that ABM systems were a destabilizing influence. Eventually, the Russians bought into this argument, and following President Nixon's election in 1969, the improved state of relations between the superpowers allowed negotiations known as the Strategic Arms Limitation Treaty (SALT) process to begin.

An unspoken reason for seeking to halt ABM systems was their high cost. Not only were the giant radars and rapid reaction anti-missile missiles very expensive, but it was always going to be cheaper for an adversary to counter these defences as explained above about the use of decoys. And you would need many radars to cover attacks from every possible angle, and many, many launchers to intercept the attacking missiles as they approach any and every part of the country. The size of the Soviet Union made this an even greater challenge than for the US, and for both countries the technology to detect missiles, including missiles from submarines just a few hundred kilometers off your coast, and intercept them before they hit was daunting and astronomically expensive.

In fact, the American Congress was almost certain to cancel the US ABM program—the high cost of the Vietnam War had put all military expenditure not related to that conflict under pressure. Congress was also balking at funding any new ICBMs or any new nuclear submarines. So, an agreement with the Soviets that had them also halt the expansion of their ABM systems, and their missile and submarine fleets seemed like a good idea.

In the ABM negotiations, the Soviets showed themselves willing to compromise by allowing the US to keep the ABM radar system around its mid-west ICBM sites, despite general agreement that it was inherently destabilizing. In return, they had the right to build one such radar system around one of their ICBM complexes. The reverse arrangement also prevailed in that the Soviets kept the ABM system around Moscow and the US had the right to build a similar system around Washington DC. Both countries also could keep the radars and rapid reaction launchers attached to their missile test ranges—the US ones were at White Sands, New Mexico, and at Kwajalein Atoll in the Marshall Islands far out in the Pacific Ocean beyond Hawaii, while the Soviet ABM test range was near Sary Shagan in Kazakhstan.

Two years after the ABM Treaty was agreed, both sides agreed to forego the right to a second ABM site and limited themselves to just the one that already existed, thus demonstrating that they had never intended to expand their ABM systems, and had agreed to do what they had intended to do anyway.

HALTING THE INCREASE IN ICBMS

As well as banning ABM systems, US and Soviet negotiators explored ways to agree a limit on *all* strategic weapons. On the face of it the list of such weapons included not only ICBMs and submarines carrying nuclear missiles, but also long-range bombers and long-range cruise missiles. A limit on the number of ICBMs also raised the question about whether they carried just one warhead or many—MIRVs.

The details of these negotiations then revolved around what to limit. Did you limit missiles, or their warheads or their silos? Did you limit submarines or the missiles they carried? Likewise, was it bomber aircraft or the cruise missiles they carried?

And what about future developments? If you banned the introduction of new missiles, when did a minor tweak to an existing missile become a new one? After 20 minor tweaks, the resulting missile would probably bear little resemblance to the one limited in the Agreement.

And when is a weapon *strategic*. The US proposed that this covered weapons that could reach one country from the other (from the northeastern border of the continental US to the northwestern border of the continental Soviet Union to be more precise) or from submarines. The Soviets wanted to include any weapons that could reach it. The difference here was that the US definition excluded the short- and medium-range missiles it had in Western Europe aimed at the Soviet Union, while the Soviet definition included them—while cleverly excluding the Soviet's own intermediate-range missiles that could hit Western Europe but not the US. Also, if a medium-range bomber can be refueled in mid-air, that may enable it to fly all the way from the Soviet Union to the US so shouldn't it be treated as *strategic*?

These and similar issues were the subject of long and detailed negotiations between the superpowers. They were also hotly debated inside the USA, with the various interested parties arguing for their particular future system to be excluded from any limits that were agreed. Henry Kissinger describes the Machiavellian and at times surreal nature of the negotiations, with the US proposing and holding out to be allowed to do what they anyway planned to do (like retire some of the old submarine-launched missiles).

The underlying consideration was whether you could check later on if the other side were cheating. Eventually, anything that couldn't be verified was omitted. Both countries recognized that SALT-I was just a temporary arrangement—the word "interim" was in its title and it had just a five-year validity. The superpowers wanted to freeze the existing strategic forces for a few years to give themselves time to negotiate something more permanent. The various topics on which agreement was not reached were added to the agenda for the next round of talks.

One of the more surreal elements that Kissinger describes was the reaction of Soviet Deputy Premier Smirnov, who ran all the Soviet defense industries, every time Kissinger mentioned the details of Soviet military strength. The CORONA and other spy satellites had given the US detailed and accurate values for the Soviet forces, so he was able to trot out figures for the numbers of each of the types of ICBM, missile silos, test flights, nuclear bomb tests, etc. Each time Smirnov reacted with fury which eventually turned to a frenzy. At that point Soviet Foreign Minister Gromyko had to take Smirnov out of the room to calm him down. Kissinger wasn't sure if Smirnov's irritation was because the US knew so many details about Soviet weapons programs or that his own colleagues (including probably Gromyko who was not yet a member of the Politburo) were learning so many details for the first time.[129] To what extent Kissinger allowed his puckish sense of humor to aggravate this situation is not recorded.

The bottom line again and again came back to whether violations by the other side of a proposed limit could be verified.

After tortuous negotiations, two agreements were signed by President Nixon and Premier Brezhnev in Moscow on May 27th 1972—one limiting ABM systems, the other holding strategic missile levels at their current levels. An ongoing process of future negotiations to reduce the number of strategic weapons was also set in train. The second of the two agreements signed that day is known as SALT-I, but in practice the whole package of agreements is loosely referred to by that name.

There was opposition to SALT-I in both the USA and the Soviet Union. The American "hawks" argued that the Soviets couldn't be trusted and that breaches of the Treaty couldn't be detected. They foresaw a future in which the Soviets surreptitiously deployed new and enhanced missiles with bigger, more accurate, and more numerous warheads.

---

[129] Kissinger (1979) p. 1234.

The American hawks pointed to the unilateral resumption of atmospheric nuclear testing by the Soviets ten years earlier in violation (they argued) of the agreement to halt such tests, and the continuing high level of tension caused especially by the conflict in Vietnam. The Soviets rebutted the charge of breaking the atmospheric test agreement, arguing that the Western allies were first to break the agreement when France exploded its first nuclear bomb in 1960.

The Soviet hawks were conscious of the suffering inflicted on them during World War II by America's new best friend in Europe—Germany. They were disturbed by the build-up of weaponry in Western Europe pointed almost exclusively at the Soviet Union, and were conscious of the superior technical prowess of the West, illustrated dramatically by the successful Apollo landings on the moon from 1969 onwards.

The political leaders in both countries therefore deserved considerable credit for resisting the pressure from their respective hawkish stakeholders.

The final text of SALT-I was relatively simple—each side would freeze its missile fleet at its current size. But the detailed negotiations behind these simple statements were fiendishly complicated. Broadly speaking, the USA kept its lead in MIRVs while the Soviets retained a lead in number and size of missiles, and in the number of submarines. Long-range bombers and cruise missiles were not even mentioned. Mobile ICBMs were included ambiguously—provided they weren't "operational" they could be deployed. "Significant" upgrades to missiles were forbidden.

The Treaty also prevented either side from concealing its activities from the other side's "national technical means of verification". This meant allowing the other side to listen in to the radio broadcasts from missile tests and not covering missile silos or submarine berths to hide them from spy satellites.

Some of the agreements were virtually impossible to verify with total assurance. Had a missile in a silo been upgraded for MIRVs? How many missiles were in the storage sheds next to the silos? But the negotiators felt that reasonable estimates could be made for these, so that the overall situation would be pretty well understood. Upgrading a missile to incorporate MIRVs was not a trivial undertaking, effectively requiring the whole missile to be replaced—and spy satellites would spot most of those occurrences. Missiles hidden in the storage sheds had to reach the site by rail, and spy satellites would spot many of these trains somewhere in the long journey from factory to site.

The number of submarines could also be monitored while they were being built at one of the Soviet Union's two shipyards equipped to build nuclear-powered subs. Each one took of the order of 6 months to fit out and was thus almost sure to be photographed at least once during its construction.

Much heartache went into whether satellites could reliably detect a "significant" increase in the size of the missile silos. The parties eventually agreed that anything more than 15% was to be considered "significant"—but whether in diameter or volume or both was left vague. This part of the Treaty has given rise to much conjecture about the resolution of America's spy satellites at that time. The nub of the matter was to detect any attempt to widen the silos of the SS-11 ICBM which was 2 meters wide to accommodate the 3-meter SS-9. On the face of it the Treaty therefore calls for the USA to detect a 30 cm widening of the silo (15% of 2 meters).

Simplistically, that suggests that the spy satellites had a resolution much better than 30 cm. In practice, such a change in width can be detected with a camera offering 50 cm or even worse resolution—the silo width can be derived from the information in several pixels to give an estimate of its width that will be accurate to much better than the individual resolution of a single pixel.

The USA only had one type of missile, the Minuteman, and had no plans to introduce a new one, so agreeing to what was already the preferred policy was a bit of a no-brainer.

When SALT-I was signed the Soviets were deploying 200 ICBMs and SLBMs[130] and launching 8 new nuclear submarines a year. The USA had stopped deploying ICBMs in 1967 and would not deploy any new nuclear submarines for another 7 years. Furthermore, the US Congress seemed set to cancel the US ABM program anyway. All in all then, the USA seemed to come out pretty well from the Treaty. In return, the Soviets ended up with a marked superiority in the number of submarines—62 nuclear subs with up to 950 SLBMs as compared with America's 44 nuclear submarines with up to 710 SLBMs.[131]

Despite the positive spin put on the Treaty by the American Administration, hawks in the USA decried the numerical inferiority of American missiles enshrined in SALT-I. At the time the Treaty was signed the United States had 1,054 operational land-based ICBMs, and none under construction, while the Soviet Union had an estimated 1,618 operational and under construction. However, the USA was in the middle of an extensive program to enhance its ICBMs and SLBMs with MIRVs, and these upgrades were permitted by the Treaty. The precision of the count of Soviet missiles was a testament to the effectiveness of the KH-5 and KH-9 satellites.

One of the more subtle outcomes of the SALT-I accords was that they legitimized satellite reconnaissance as a means of verifying compliance with the agreements.[132]

[130] Kissinger (1979) p. 821.
[131] "... United States and its NATO allies have up to 50 such submarines with a total of up to 800 ballistic missile launchers thereon ..."—this SALT-I statement of May 17th 1972 by the Soviet side includes the British and French forces.
[132] Gaddis (2005) p. 200.

7

SALT-II

Neither side was terribly happy with the SALT-I agreement, which suggests that it was fairly equitable. It was intended as a stop-gap measure designed to give the superpowers a chance to negotiate a more permanent limit on nuclear weapons. It was thought that negotiations to reduce weapon levels would be easier in the knowledge that the threat was not growing rapidly.

That was the theory. In practice, the arsenal of both the US and the Soviet Union *was* growing rapidly despite SALT-I.

In this chapter we run a check on the nuclear firepower of the superpowers in the 1970s. Then we look at the main issues bedevilling negotiations of further arms agreements including bombers, cruise missiles, MIRVs, and more. The verification of arms limits and the SALT-II itself are then discussed—the verification dictating what the Treaty could realistically contain. A final summing up looks at the significance of the SALT-I and SALT-II agreements, even though they ended with a bit of a damp squib.

HOW MANY IS ENOUGH?

The US should have felt stronger after SALT-I by virtue of its ongoing program of fitting its missiles with Multiple Independently-targeted Reentry Vehicles (MIRVs) and other new developments. The US lead in miniaturization and electronics gave them a lead of five or more years over the Soviets in the MIRV game—the first Soviet test of a MIRV-ed missile didn't take place until August 1973,[133] after SALT-I was signed. But the US hawks saw it differently—the Soviets were allowed more missiles under SALT-I, some of which were much bigger than American missiles, so sometime in the future they could pack them with lots of MIRVs.

---

[133] York (1973) p. 18.

The limits set by SALT-I allowed the Soviets to keep the 1,618 ICBMs they had already deployed or were in the process of doing so, while the US had to stick with its 1,054—made up of 1,000 Minuteman and 54 older Titans. Not only did the Soviets have more missiles, they were bigger ones. Included in the Soviet ICBM force were the large SS-9 missile and the SS-18 that would replace it. Their massive nuclear warheads were seen as capable of destroying even the most deeply buried missile silos or command centers. In terms of the weight of the warheads that Soviet ICBMs could deliver, they had roughly twice the total capacity of the US ICBM fleet.

SALT-I also allowed the Soviets to have more submarines and associated missiles than the US. They were allowed up to 62 submarines compared with 44 for the US, and 950 missiles as opposed to 710 for the US.

The Soviets may have had the edge in raw missile firepower, but the US could destroy more targets. Many of the US missiles were fitted with MIRVs and the rest were in the process of being upgraded. Each MIRV warhead represented a separate Soviet target that could be wiped out—airfield, missile silo, factory, harbor, power station, railway marshalling yard, military base, etc. Each MIRV warhead had "only" about a quarter of a megaton of killing power in the case of land-based ICBMs and 50 kiltons in the case of SLBMs, but that was more than 10 times (for ICBMs) or 3 times (for SLBMs) the power of the bombs that destroyed Hiroshima and Nagasaki. So in terms of practical military impact, the US arsenal in 1971 while SALT-I was being negotiated contained about 4,000 warheads, almost double that of the Soviets.[134] By the end of the 1970s—when you counted in the completion of the replacement of 550 of the Minuteman II ICBMs with the 3-MIRV Minuteman III, and of 500 of the Polaris SLBMs with the 10–14 MIRV Poseidons—the number of US warheads approached 9,000 while the Soviet count grew but was fewer than 3,000.[135]

And the US had an edge in bomber aircraft. The 150 or so Soviet Bison and Bear long-range bombers were far outweighed by the 500+ American B-52 (recently refurbished) and FB111 bombers. And the supersonic B-1 strategic bomber was under development, and the stealthy B-2 was in the development pipeline after that.

One reason the US hawks weren't mollified by having more warheads than the Soviets was that they were still reeling from the speed with which the Soviets had recovered from having barely a handful of ICBMs in 1961 to having caught up with, and surpassed, the US numbers. During the mid- and late-1960s the Soviets were deploying 200 land-based ICBMs a year—building the silos, the railway tracks, and roads for access, the protective anti-aircraft batteries, the electricity plants, the communications connections, and all the other facilities to turn a barren piece of remote steppe or forest into a strategic weapon.

The Soviets had also built and deployed missile-bearing submarines at a rate of knots. By 1966, the US had deployed (at sea and operational) 41 nuclear-powered Polaris submarines each carrying 16 missiles with a range of 2,200 km (most were upgraded later to 4,500 km), which alone provided more than enough of a strategic

---

[134] Scoville (1971) p. 20.
[135] Scoville (1972) p. 15, and Kissinger (1999) p. 119.

Figure 35. A Polaris submarine-launched ballistic missile (SLBM) breaks the surface having been launched underwater.

deterrent—never mind the 1,000 Minuteman missiles and the squadrons of B-52 bombers. That same year (1966), the Soviets launched their first nuclear-powered missile-launching submarine, illustrating that they lagged the US in the SLBM department by at least 5 years. SLBMs are stored upright inside a submarine, so they tend to look fatter and squatter than the long slim ICBMs. They are launched while the submarine is underwater—compressed air squirts them to the surface whereupon their rocket motor ignites (Figure 35) Longer range missiles give the submarine more of the ocean to operate in—they can hit a target from farther out at sea. But the standing-up arrangement means you can't make the rocket longer, and making it fatter is intrinsically aerodynamically inefficient when it's in flight. So, the Soviet penchant for large rockets didn't apply when it came to the submarine variety.

By the time SALT-I was signed the Soviets were launching nine submarines a year and by the mid-1970s their fleet exceeded the size of the American one, and like the US Polaris each Soviet submarine held 16 missiles. However, that apparent Soviet advantage overlooked the lack of MIRVs—almost all of the US SLBMs were MIRVs, while none of the Soviets' was. And the Soviet missiles had a range of only 3,000 km, so they had to get fairly close to the US coastline to be able to hit their targets. Not only did that make them vulnerable to being located by the US submarine detection networks, it also meant that they had to sail for several days to reach their on-station position, and the same again when returning to base. The US could base its submarines in friendly sites in the UK, Spain, and the Marianas Islands close to the points where they could hit their targets inside the Soviet Union, thus requiring fewer days at sea. This geographical advantage converts into more submarines on station given that for every two months at sea, one month has to be spent back at base for refurbishment and crew switch-over. Of the two months at sea, American submarines spent more time on station than did Soviet ones because their bases were closer to hand. This fact alone negated a good part of the apparent Soviet numerical advantage.

So, SALT-I elicited compromises from both sides, and represented a reasonably good starting point for the tougher negotiations in SALT-II covering the complete range of strategic forces.

SALT-I had deliberately ignored some of the more intractable issues. Long-range bombers such as America's B-52s were not mentioned and neither were cruise missiles. These were two of the issues that critics of a new agreement considered difficult to verify. Other such issues included the number of MIRV-ed vehicles, the upgrade of ICBM systems, the number of "rapid reload" spare ICBMs at a site, and the number of mobile ICBMs.

Let's have a look at some of the complexities of these issues as a way to understand why getting this sort of agreement was so difficult.

BOMBERS

By any standard, the American B-52 bomber was a successful design. First flown in 1952 and last manufactured in 1962, it is expected to continue in service to 2040, albeit with many modifications. For the purposes of SALT negotiations, the B-52 was considered to be a H-bomb dispensing machine, either directly over the target in the way the B-29 bombed Hiroshima and Nagasaki, or indirectly via an air-launched missile (ballistic or cruise). In fact, the B-52 could dispense all sorts of munitions, and had been doing so over Vietnam and Cambodia for some time. Twenty years later it was still the work-horse of the US bombing fleet, dropping 31% of the bombs in the 1991 Kuwait conflict.[136] And a decade later which bomber was called on to bombard the Tora Bora caves and passes in Afghanistan where Osama Bin Laden was supposedly hiding? That's right: the same B-52 (see cover photo).

The US also had a hundred or so of the sophisticated, expensive, and relatively unreliable bomber version of the swing-wing F111 airplane. Built with the enthu-

---

[136] Edgerton (2006) pp. 95, 155.

siastic support of former Defense Secretary Robert McNamara, the F111 was supposed to save money by being able to accomplish more than one mission—fighter as well as bomber, Navy as well as Air Force. You know what they say: "jack of all trades, and master of none." The F111 lacked the range and the carrying power of the B-52, but it was faster. Does a bomber have to be fast? It could handle intercontinental distances with a little help from the in-air refueling boys.

The Soviets had never gone in for long-range bombers in a big way—that is to say, bombers with an intercontinental range. Their historical enemies were closer to home, and could be reached with smaller (and cheaper) bombers. Stalin and Khrushchev both liked the idea of building long-range missiles that would avoid the need for such bombers, and leap-frog the advantage that the Americans had in this department. From the 18 Bison aircraft so carefully counted by the US Air Attaché in 1955 (see Chapter 5), the fleet had grown to about 50. In addition the Soviets had about 100 of the turboprop Bear aircraft that had the power and the range to be counted as strategic.[137] The Soviets claimed that the TU-26 Backfire bomber which first flew in 1969 did not have the range to be counted as strategic, being intended to target enemies in Western Europe, China, and any maritime targets close to the Soviet borders. Hawkish, especially naval, voices in the US argued strongly that the Backfire could reach the continental USA from the Soviet Union and so should be counted—but it would have to fly at high altitude to conserve fuel (a sitting duck for US air defences), come from an Arctic airfield, and probably have to be refueled in mid-air at least once. Satellite imagery of the Backfire was evaluated by both the Defense Intelligence Agency (DIA) and the civilian CIA—they calculated the size of the aircraft and used thermal imagery to establish the fuel capacity inside its wings, and brought in aircraft designers to advise on what its range would be. That sort of evaluation is not an exact science and they came up with differing answers, although both put its range beyond the 5,500 km definition[138] of a strategic bomber. The DIA decided that the Backfire had an un-refueled range of between 8,300 and 11,000 km, and that it could be refueled in the air. The CIA put the range somewhere between 6,500 and 9,300 km, and claimed there was no sign of the appendage required for in-air refueling, and anyway the Soviets had very few long-range tankers to do the refueling. An unhappy compromise estimate of 9,600 km was agreed—the compromise was unhappy in that neither side conceded its fundamental position: the CIA that the Backfire was not a strategic delivery system, the DIA that it was.[139] Note that the US Navy (with some influence on the DIA, presumably) was very keen for the Backfire to be included in the negotiations because it was considered a serious threat to the US fleet.

An interim agreement reached in 1974 removed both the Soviet Backfire and the US F111 from the negotiations, but the US hawks refused to accept that as the final position.

---

[137] Scoville (1971), and Aspin (1979).

[138] SALT-II Article II.1.

[139] Lindgren (2000) pp. 136–137, and Burrows (1986) pp. 230–231.

CRUISE MISSILES

By the 1970s the cruise missile was beginning to have the characteristics that we have come to associate with it today—low-flying to evade radar, and precision-targeting. It was no longer the buzz bomb of World War II which was pointed in the general direction of London and dived into the ground with its bomb load when it had flown for a pre-calculated time or when its fuel ran out. The US had a big lead in cruise missile technology and had made them small enough to fit *inside* a carrier aircraft. Being inside, you couldn't ascertain anything about the nature of the cruise missile— was it long-range or did it carry a nuclear warhead, for example? The Soviet cruise missiles were still slung under the wings or belly of the carrier aircraft, and thus could in principle be seen and analysed in photographs.

But you can't really ascertain the range of a cruise missile by looking at it. A small one certainly has a short range, but a large one could have a heavy payload (i.e., a bomb) and a short range, or a not-so-heavy payload and a longer range—you could trade fuel and thus longer range for payload.

You also couldn't ascertain how many cruise missiles were inside the carrier aircraft because they were, well, inside.

The best the SALT-II negotiators could think of was to have both sides promise not to load up aircraft with more than a certain number of long-range cruise missiles, and not to have more than a certain number of those kinds of aircraft. If the Americans had wanted to cheat on this point they could load up planes with lots of cruise missiles and either not tell the Soviets, or claim they were short-range vehicles. Eventually, the Soviets would be able to cheat like that too, once they caught up with the US on miniaturization and efficiency of cruise missile technology.

Cruise missiles launched from the ground and from ships or submarines were also a thorny issue. The Soviets hadn't got any of those, so they wanted a total ban on them. The US wanted these weapons to be exempt from the Treaty in the same way as short- and medium-range missiles were. In the end the two sides couldn't agree and left this class of weapon for the next round of negotiations after SALT-II.

MULTIPLE INDEPENDENTLY-TARGETED REENTRY VEHICLES

The heart of the SALT-II negotiations concerned the number of ICBMs and SLBMs and the number of warheads they carried. The plan was to put a limit on both— missiles and warheads.

The latest types of Soviet missiles were fitted with Multiple Independently-targeted Reentry Vehicles (MIRVs) (Figure 36): the SS-17 (4 MIRVs), SS-18 (8), and SS-19 (6). Older Soviet missiles either carried a single warhead or in the case of the SS-11 and SS-N-6 (SLBM) had two or three warheads that were not independently targeted—they just hit the same target slightly apart, shot-gun fashion.[140]

[140] Aspin (1979).

Figure 36. MIRV re-entry vehicles for the US Peacekeeper ICBM.

What was to stop the Soviets replacing the single warheads with a MIRV one? In principle, this would be hard to detect if the changes were made inside a silo or submarine. However, we are not talking about taking one large suitcase out of the trunk of a car and replacing it with three or four small ones. It would be more like removing the whole trunk and replacing it with a three- or four-piece rear compartment. And the chassis and fuel tank would have to be moved to accommodate the new arrangement, otherwise the vehicle will be unstable. These considerations applied for existing ICBMs and SLBMs. The warhead and rocket were designed as one vehicle and changing one and not the other was not simple. So, in theory each side could cheat by replacing single warheads with MIRVs, but in practice this was something the older missiles had not been designed for.

Other ways to cheat on the MIRV limits would be to replace old missiles in silos or submarines with modern MIRV-ed ones, but this was reasonably straightforward to detect from space—each missile type required different support facilities, and the change-over would be a major exercise, impossible to hide on any significant scale.

OTHER ISSUES

At the start of this discussion I mentioned upgrading missiles, spare missiles, and mobile missiles as other issues that caused SALT-II negotiator heart-ache.

The simplest answer to upgrading missiles was to forbid it. In the end it was agreed that each side could introduce one, and only one, new type of missile. As discussed in Chapter 5 a new missile has to be tested many times before it is considered operational, and those tests are readily detected by a combination of satellites, ground radar, and radio-listening facilities (on ground and in space).

One type of potential Soviet cheating was difficult to monitor from space—it was theoretically possible to add an upper-stage rocket to the intermediate range SS-20 missile and turn it into the SS-16 ICBM (the SS-20 was basically the first two stages of the SS-16). The idea would be for the Soviets to stockpile lots of the upper stage and the associated warhead, then at short notice upgrade all their SS-20s to SS-16s. The agreed way round this potential loophole was for the Soviets to dismantle all their SS-16 ICBMs. Then if they did upgrade an SS-20 with an upper stage, the combination would never have been tested—and testing can be verified with a high degree of certainty.

The rapid reload of spare missiles into a silo after firing the first missile had never been tested, so was discounted as a major risk, although it came back to haunt negotiators in the 1980s.

Land-based mobile launchers (typically railroad cars) were impossible to count with any degree of certainty. The Soviets had deployed some of these already and the US was considering it for the MX missile that was planned to replace the Minuteman. One tactic would be to build thousands of shelters for the missiles and only occupy a small number of them—the "shell trick" scenario as it was called when the American MX designers were considering this option. This subject was put into the "too difficult" pile and left for the negotiations after SALT-II.

SIGNIFICANT AND HEAVY

That left a thorny issue that infuriated right-wing commentators in the US—when does a "light" launcher become "heavy", and what is a "significant" upgrade?

SALT-I had said that missile silos could not be expanded by more than 15%. The Soviets interpreted that to mean 15% in diameter *and* 15% in depth, making the silos a third bigger than before. With this liberal interpretation of the agreement the Soviets began rolling out bigger and better missiles throughout the 1970s—the SS-16 carried a warhead twice the size of the SS-13 it replaced, the SS-17 had three or four MIRVs that together were double the warhead of its SS-11 predecessor, and the SS-18 had eight or nine MIRVs giving it an even larger warhead capability than the massive SS-9 it replaced. These two new MIRV-ed missiles achieved some of their enhanced capability by "cold launching"—popping the missile up out of the silo using an auxiliary engine then firing its main rocket engine. This technique not only increased the missile's carrying capacity but had the worrying advantage that the silo could be reused. The fourth of the new Soviet missiles, the SS-19, was larger than the SS-17 although not as huge as the top-of-the-range SS-18.[141] To American eyes this

---

[141] Lindgren (2000) pp. 132–133, and Kissinger (1982) p. 1011.

across-the-board enhanced ICBM fleet seemed to make a mockery of SALT-I's "no significant" (i.e., 15%) upgrade to the silos.

What was interesting of course was that the US was able to detect this dubious interpretation of the 15% limit. It showed both the capability of spy satellites to monitor such details with considerable precision and the need to tie down agreements unambiguously.

The Soviet and American understandings of the phrase "heavy launcher" were also different. SALT-I prohibited the upgrade of existing "light" launchers to the "heavy" category. You could "modernize" them but only within a silo not "significantly" (greater than 15%) enlarged. Henry Kissinger made the US view of this clear by making a formal statement that the US considered anything larger than the SS-11 to be "heavy". The Soviets said "nyet" to that, and took the word "heavy" to mean the top-of-the-range SS-9 or bigger. Hence the various older missiles mentioned above (SS-11, SS-13) were "modernized" without becoming "heavy" within the Soviet meaning of these terms. This rankled with the US, especially in the case of the SS-19 which was considered by the Soviets as a "light" ICBM but its six MIRVs contrasted with the three usually carried by the US Minuteman (although it had a theoretical maximum of seven).

Of course, the US wasn't standing still during the 1970s. As mentioned above, among the new weapon systems being developed were a new generation of cruise missiles, the Trident submarine and its SLBM which had begun test launches in 1978, and the B-1 bomber.

SETTING LIMITS—AND VERIFYING THEM

By 1974 the Soviets had pretty much deployed as many missiles as they wanted. Henceforth they would focus on the upgrades mentioned above for both land-based and submarine-based missiles. This left them ready and willing to make some kind of agreement.

President Nixon had hoped to complete a new agreement before he retired in 1976 but his premature resignation in August 1974 over the Watergate affair prevented that. Nevertheless, his successor, Gerald Ford, with Henry Kissinger retaining his role as Secretary of State, picked up the baton and reached an important agreement in December that same year (Figure 37). In a deal usually referred to by the name of the venue where President Ford and Soviet leader Brezhnev met, Vladivostok, the superpowers agreed to put a cap of 2,400 on the number of delivery vehicles (missiles plus long-range bombers) and within that a limit of 1,320 on the number of those that were MIRV-ed. The diplomats were then tasked with turning that into a water-tight Treaty.

In 1979, just before the resulting SALT-II Treaty was signed by President Carter and Premier Brezhnev, US Congressman Les Aspin summarized the verification issues.[142] Aspin was recognized as an expert in defense matters at the time, and

---

[142] Aspin (1979).

Figure 37. President Ford (left) and Premier Brezhnev sign the Vladivostok Agreement, November 24th 1974.

six years later became Chairman of the House Armed Services Committee, and six years after that became President Clinton's Defense Secretary. His analysis was considered authoritative at the time, because he was assumed to have access to much of the classified information on spy satellites, radars, and other verification techniques and technology.

Adequate verification of compliance was the *keystone* to SALT-II according to Aspin, and he then analysed how the USA would be able to verify the various forms of cheating that the Soviets might attempt. Surveillance satellites were central to detecting most of the cheating as his analysis showed:

- The limits set on the number of bombers was easy to monitor since the Soviets only had about 10 airfields able to handle heavy bombers.
- The limit on the number of SLBMs was monitored by satellite images of the construction of new submarines or the enhancement of old ones both requiring many months or even years in the two or three suitably equipped Soviet shipyards, and then extensive sea trials that could be followed using radio intercept and sonar.
- The number of ICBMs was monitored by satellite imagery: the high-resolution satellites could identify the type of missile, and multi-spectral satellite images could detect underground and night-time deployment. It was accepted that small-scale violations would be possible if only because of the time taken to process images and analyse them.
- The substitution of MIRV-ed missiles for non-MIRV missiles was verifiable by satellite because the missiles themselves had to undergo significant changes to

Figure 38. A Tomahawk Block IV ship- and submarine-launched cruise missile is trailed by a Navy F-14 Tomcat during a test flight. Credit: US Navy.

carry MIRV warheads. SLBMs with MIRV warheads need larger tubes in the submarine which again satellite images can spot.

- Cruise missiles on aircraft could be counted by satellite imagery if mounted on the outside of the aircraft which all Soviet ones at the time were.
- Substitution of big missiles for smaller ones could be detected by satellite imagery—the missiles might be wider or longer, both of which could be measured when they were being tested.
- Mobile ICBMs were impossible to count accurately. The only verifiable option was to ban this type of weapon altogether—even testing of them was forbidden, since once they had been successfully tested they could perhaps be deployed surreptitiously.
- Long-range cruise missiles launched from land or sea were forbidden for the same reason that their numbers weren't verifiable (Figure 38). Testing of them was also prohibited which effectively barred the Soviets from catching up with the US—Aspin estimated they were 10 years behind America in this type of weapon.
- Several other forms of potential cheating were verified using techniques other than satellite images. Upgrades of the Backfire bomber to give it a longer range would be detected by eavesdropping on air-to-air radio links during in-the-air refueling. Ground-based radar tracking and radio intercept of telemetry during

testing would pick up attempts to add a third stage to the SS-20 and thereby make it into an SS-16, to replace a single warhead with a MIRV-ed one, or to try out a rapid reload system, since all of these would first need to be tested.

Aspin noted however that a number of potential infringements could not be reliably verified and these were left for future negotiations. He listed three of them: the range of cruise missiles, carrying cruise missiles inside an aircraft instead of slung beneath the wings, and the possible design of a MIRV warhead that could be easily transferred to what was previously a single-warhead missile.

Several Soviet violations of the SALT-I agreements were detected during the period of the SALT-II negotiations, and the Administration tried to keep these from the Congress in order to avoid a backlash against further negotiations.

One clear example came when the Soviets launched four new nuclear submarines in 1975 whose missiles would take the Soviet total over the limits set in SALT-I. They were therefore required under the Treaty to destroy an equivalent number of older missiles, but they failed to do this in time. The matter was raised by the US at the official meetings and the Soviets admitted that they were in breach of the agreement and duly completed the destruction of the missiles.[143]

Many commentators felt that the verification of the limits being discussed for SALT-II was sufficiently fool-proof that the Treaty should be approved. But the US system requires not that the majority of the decision makers agree, but that two-thirds of them do. The Founding Fathers wanted the New Republic to avoid the kind of "foreign entanglements" that the European states tended to engage in, and wrote the Constitution to make it difficult for foreign treaties to get Congressional approval. The need for a Senate two-thirds majority to approve Treaties is the mechanism that enshrines that intention of the Constitution writers—and it works.

One of the consequences of needing a two-thirds majority in the Senate for Treaty approval is that the negotiators end up trying to fine-tune the content of the Treaty to influence the vote of two or three Senators in order to get the required majority. Even if a Treaty has broad support, as was the case with SALT-II, it is difficult to get above the magic 66% mark. Many Senators were philosophically against voting for a Treaty with a communist dictatorship, no matter how beneficial the Treaty was for the US. Others deemed their political future to be at risk if they supported it because of the anti-communist leanings of their particular electorate. Yet others would have tactical reasons for opposing the vote, perhaps looking for support for one of their favored bills in exchange.

The Soviet deployments could be counted by the spy satellites, but their intentions could not. The DoD analysts tended to see future threats where the civilian CIA analysts didn't. For the most part the Nixon, Ford, and Carter Administrations were not unduly influenced by the more alarmist of these interpretations, but there were always a few Senators and Congressmen who were. Leaks to Congress or influential media personalities brought these disagreements into open debate from time to time. An example often quoted is the interpretation of the huge Soviet investment in civil

[143] Lindgren (2000) p. 138.

protection facilities. Details of the scale of this varied widely depending on which source you referenced but at its most extreme it appeared that over 100,000 workers were involved in building underground facilities in Moscow and other major cities—enough for perhaps 20% of the total population. Critics saw this program as part of a Soviet plan to launch an attack against the US while sheltering the Soviet population. Others saw it as a Soviet over-reaction to the US policy of Mutually Assured Destruction (MAD) with no real implication for US security. The matter was never resolved satisfactorily, and still remains something of a mystery.[144]

THE SALT-II AGREEMENT

Despite the difficulties only really hinted at above, the US and Soviet negotiations did reach agreement on what became known as SALT-II in early 1979, and it was signed by President Carter and Premier Brezhnev in Vienna in June that year (Figure 39).

The deal reached in Vladivostok four and a half years earlier was broadly maintained, including the overall limit of 2,400 "nuclear strategic delivery vehicles" (missiles and bombers), which was to be reduced in 1981 to 2,250.

Within that overall deal was a series of nested limits for various forms of "delivery vehicle" and warhead. No more than 1,320 of the 2,400 could be MIRV-ed launchers or bombers with cruise missiles. No more than 1,200 of that 1,320 could be MIRV-ed ballistic missiles (i.e., rockets as opposed to *cruise* missiles). And of that 1,200 no more than 820 could be ICBMs (the other 380 were presumably SLBMs).

Each side was allowed to introduce one new type of missile.

To make all of this water-tight, the agreement contains extensive language to ban all sorts of feared violations such as improving an existing missile so that it effectively becomes a new one, limiting the number of cruise missiles in an aircraft, and the dismantling of the Soviet SS-16 missiles (because of the fear of the SS-20 being surreptitiously upgraded to effectively become SS-16s).

The agreement does not put an explicit limit on the number of warheads, but it left intact the US arsenal of about 9,000 warheads (it depends how many cruise missiles you count) and a Soviet total of about 5,000. Both figures are of course a grotesque over-kill whether for deterrence or attack.

Having been signed by President Carter the Treaty was then forwarded to the Senate in the usual way for its "advice and consent".

SALT-II came to the Congress for discussion at a bad time. There was a long-standing criticism of the Soviet human rights record, especially concerning the Soviet refusal to allow Jews to emigrate to Israel. You will recall that the Soviets had to build a barbed wire barrier (a wall in Berlin) across central Europe and shoot any of its citizens who tried to cross it. They did issue some exit visas each year for Israel, but nothing like enough to satisfy the demand of Soviet Jews. The issue became a Congressional matter because the Administration proposed to give the Soviet Union

[144] Lindgren (2000) p. 141, and Burrows (1986) pp. 4–8.

Figure 39. President Carter and Premier Brezhnev signing the SALT-II Agreement at the Kremlin, June 1979.

so-called Most Favored Nation (MFN) status—which actually means nothing more than no longer being "un-favored", thereby joining the 100+ countries that the US trades with routinely. But the attempt to upgrade the Soviet Union to this benign category of trading partner was a trigger for the emergence of long-standing Congressional criticisms of the Soviets, high among which was the Jewish emigration issue, resulting in the Jackson–Vanik amendment barring MFN status to countries with emigration restrictions.[145] The Congressional action made matters worse for Soviet Jews—while the US legal action was threatened the Soviets opened their borders to some extent, but once the law was passed they clamped down again (30,000 exits a year before passage of the Amendment; 15,000 after). The criticism rumbled on.

It got worse: in December 1979 the Soviets invaded Afghanistan making any agreement with them anathema to the American Congress and public. In addition, the authority of the Carter Administration had been severely weakened by the crisis over the American Embassy staff in Tehran being taken hostage in November 1978 by the revolutionary forces that took over Iran and ousted the Shah.

Afghanistan proved to be the last straw and the Carter Administration felt compelled to withdraw SALT-II from Congress. It was left for the more hawkish Republican Administration of President Ronald Reagan to pick up the subject of arms negotiations in a new and aggressive way in 1981.

[145] Kissinger (1982) p. 986.

SUMMARY

SALT-I and SALT-II were far from perfect, but they showed that there was a lot of common ground between the two "warring" superpowers. And they showed that with adequate verification, limits could be put on military weaponry and the escalating nuclear arms race could be slowed down. Neither side was going to trust the other, not when their vassal states were engaged in wars as violent as Vietnam. It was in their mutual interest to stop the expensive growth in nuclear weapons and their delivery vehicles since that growth gave neither side a significant advantage. Furthermore, these same nuclear arsenals were of no practical use in the myriad of regional conflicts like Afghanistan and Vietnam. Both superpowers had to accept that they were vulnerable and that this was a form of security—it assured the other side that you would refrain from aggression, which thereby reduced your risk of being attacked. As Henry Kissinger put it "For the first time in history two major powers deliberately rested their security on each other's vulnerability,"[146] but he also noted that "each of the provisions of the basic accord required detailed verification procedures."[147]

[146] Kissinger (1982) p. 256.
[147] Kissinger (1999) p. 848

8

The other Cold War nuclear powers—China, the UK, France

The three minor countries with nuclear weaponry during the 1960s were China, Britain, and France, and their contribution to the story will be told in this chapter. India joined the "nuclear club" in 1974, but as part of the non-aligned community of countries did not have a direct influence on the superpower dialogue, so its story will be covered in Chapter 9.

During the Cold War, the nuclear forces of Britain and France were targeted at the Soviet Union and its allies. SALT-I and SALT-II were able to ignore the French and British nuclear arsenals because they paled into insignificance in comparison with the vast number of American missiles with which the Soviets understandably considered Britain and France to be aligned. The combined French and British strategic missiles at the time of SALT-I, for example, were less than 5% of the American and Soviet aggregate total, and the proportion went down as the superpower arsenals expanded rapidly after that—Britain's missile count was less than 1% of the total by 1977. Why did they feel it necessary to possess these expensive weapons of mass destruction despite the increasing awareness of their limited military utility?

The Chinese nuclear arsenal was a deterrent against primarily the USA but also to some extent against the Soviet Union and Japan.

BRITAIN

Immediately after World War II Britain was effectively bankrupt and sought any measure possible to reduce government expenditures without compromising its imperial ambitions across the globe. Britain had collaborated with the USA in the development of the atomic bomb, merging its research activities into the Manhattan Project that developed the bombs dropped on Hiroshima and Nagasaki that ended the War. The two countries had negotiated an exclusive right to import uranium from the Belgian Congo (with half paid by Britain) and agreed to divert all of that to the

US for the duration of the war. Wartime agreements signed in 1943 and 1944 gave the UK and Canada a veto over US use of the atomic bomb and gave Britain joint rights with the US for developing it after the War. These agreements were in conflict with the Atomic Energy Act passed in 1946 which prohibited sharing design information with foreign countries—an anomaly that Congress was not aware of when the Act was passed, and which the Administration therefore needed to resolve.

The Manhattan Project had relied heavily on Congo uranium, so after the war when Britain started receiving the half of the ore for which she was paying, the effect on the US was immediate. Britain had no urgent use for the uranium and was storing it for a time when Britain's own bomb had been developed. The chance for the US to rid itself of the British atomic veto and uranium dependence came in 1946–1947 when Britain came cap in hand looking for an additional loan—the US was already propping up the shaky UK economy through extensive loans granted during the War. The UK negotiators, particularly the economist John Maynard Keynes, assumed that the US would feel obligated to Britain for its steadfastness throughout the War. They got a rude awakening. By the time they got an extra loan, Britain had been forced to give up any claim on US nuclear technology, transferred to the USA two-thirds of the Congo uranium obtained since the end of the War, and diverted all Congo uranium to the US for the following two years. In addition to the desperately needed loan, the UK got access to some information on nuclear reactors for electricity generation.[148]

The atomic bomb was widely perceived to have brought a rapid end to the War in the Pacific, and this apparent fact was not lost on British policy makers. Indeed, it seemed in the late 1940s that possession of an atomic bomb was a relatively cheap way to exert military pressure anywhere in the world. The decision to build a British bomb was therefore taken with the objective of reducing conventional military forces and thereby reducing government defence spending. You may recall from Chapter 3 that President Eisenhower started out with the same idea.

Of course, savings didn't happen like that in practice. Yes, the British nuclear bomb did probably act as a deterrent to Soviet invasion of Western Europe and probably cost less than the enormous conventional armed forces that would have been needed to achieve the same level of deterrence. But the public horror at the use of nuclear weapons to kill civilians is so strong that politicians have balked at using them ever since World War II. So, nuclear weapons have proved useless in the many regional wars across the globe since World War II, requiring countries involved in those wars to field the same large conventional forces that were required before the nuclear age. And Britain was involved in a variety of small-scale military actions throughout the 1950s and 1960s: in Malaysia against communist rebels, and in Kenya and Cyprus against anti-colonial fighters, for example. The most ill-conceived and ill-fated of these military adventures was the 1956 Anglo-French invasion of Egypt to prevent the Suez Canal being nationalized by that country's Arab nationalist regime led by the charismatic President Nasser. Executed with the surreptitious but active aid of Israel, the Anglo-French invasion was a military success but a political disaster—

[148] Rhodes (1995) p. 300.

it did for Arab relations with Britain and France what the 2003 invasion of Iraq has done for Islamic relations with the US (and Britain, again)—that is, soured them for years to come. President Eisenhower was appalled by the imperialist arrogance of the Anglo-French attack and used America's economic and political power to force their withdrawal from Egypt. The atom bomb was no help in the Suez adventure, nor in any other of Britain's military engagements of that period.

But despite the elusiveness of defence savings, once Britain started down the nuclear road it proved hard to turn back.

The "save money" argument for possessing the bomb was gradually forgotten, and replaced by two other rationales—one concrete and one fuzzy. The concrete argument was to respond to the perceived Soviet threat. Russia had exploded its first atomic bomb in 1949, and had shown other aggressive tendencies such as blockading Berlin, and refusing to hold free elections in any of the Eastern European countries it occupied at the end of the War. The NATO alliance tying the USA to defend Western Europe from Soviet attack only came into existence in 1949 and took some time to establish its credibility, so a British nuclear weapon seemed a sensible defensive move until at least the late 1950s, allowing the UK to deter the Soviets with the threat of an unacceptably ruinous British retaliation to a Soviet attack—and without the need to drag US nuclear weapons into a Soviet–British confrontation.

The fuzzy argument for possessing the bomb is the view that to participate at the "top table" of international negotiations it is necessary to possess a nuclear weapon. A variant of this argument is that to retain its Permanent Seat in the UN Security Council a country is expected to be a nuclear power. Neither of these versions of the argument is in any way official, but each is deployed from time to time to explain Britain's need to retain a nuclear capability. Bizarrely, this argument implies that Britain is trying to impress its main ally, the US, as much as it is trying to scare its enemies.[149] For example, Britain is said to have played a major role in the test-ban negotiations of the early 1960s and in the non-proliferation treaty negotiations at the end of that decade by virtue of its nuclear status.[150] This argument is a dangerous one in that if top-level political influence requires a country to possess the bomb, a country seeking such influence will be motivated to acquire a nuclear arsenal. However, the evidence for this argument is pretty thin—if you need an atom bomb to be an international heavyweight why have Japan and Germany, for example, been so influential since the 1960s, since neither possesses the bomb? Those two countries show that economic strength is at least as powerful a reason for being listened to as nuclear weapons. China's recent increasingly influential involvement in world affairs has little to do with its nuclear deterrent and all to do with its economic growth.

The close ties between America and Britain in nuclear matters during the Manhattan Project were re-established after Britain's first atomic bomb test in late 1952—and continue to this day. Britain went on to develop a H-bomb, although the first two tests in 1957 were only partially successful. In total, the UK has conducted 45 nuclear tests, all since 1958 underground, the most recent in 1991.

---

[149] DeGroot (2005) pp. 221–222.
[150] Johnson (1983) p. 467.

When it was involved in atmospheric testing, Britain made sure to locate them about as far as it is possible to get from Britain—in Australia and the South Pacific: nine in the State of South Australia, three on the Monte Bello Islands off Australia's northwest coast, three on Malden Island, and six on Christmas Island, both in Kiribati a former British colony (Gilbert and Ellice Islands) in Polynesia—the US conducted 24 tests on Christmas Island some years later. Further British tests were underground—under America that is, not Britain: 24 tests in Nevada from 1962 to 1991.[151]

The British delivery platform in the 1950s was its squadrons of bombers, but by the early 1960s it was clear that this form of nuclear delivery was behind the times. In its efforts to keep up with the changing nature of the Soviet threat, it wasn't long before Britain found that the price of remaining in the nuclear club was beyond its means. The maximum warning time (from radar) of an incoming Soviet missile aimed at Britain was only about 5 minutes. Aircraft on the ground would have no chance to get aloft before such a missile arrived and by the early 1960s it was clear that the Soviets would soon have more than enough missiles to knock out all potential British missiles based on land—a British development program had created a reliable long-range missile that could have formed the basis for a land-based missile force. Britain lacked both the vast land area of the USA in which to base thousands of missile silos, and the extra 10 minutes of warning time due to geography that made it just feasible for American aircraft to scramble into the air before a horde of Soviet missiles arrived. It was recognized that the only viable form of retaliatory British missile was submarine-based—once at sea, submarines were relatively undetectable and therefore invulnerable to a surprise Soviet attack.

Having aimed most of its research into bigger and better bomber aircraft and little or none into submarine-launched missiles, Britain was forced once again to go cap in hand to America in 1962 to keep its nuclear deterrent credible. In an arrangement that still stands, America provided Britain with the missiles (Polaris, and later Trident) to be carried in British submarines and armed with British nuclear weapons.

That decision effectively ended a stream of UK missile development that derived from V-2 missiles captured from Germany by the British at the end of World War II. The UK had improved the technology with some help from US industry and produced a reliable and powerful rocket called Blue Streak. However, Blue Streak used a volatile liquid fuel, hydrogen peroxide, which made it complicated and time-consuming to prepare for launch, and that was not the basis for a fast reaction nuclear-tipped missile. In 1960 the UK government was faced with the decision of redeveloping Blue Streak to make it an effective military weapon or to buy an American alternative, and in April that year they canceled Blue Streak and chose an air-launched American missile called Skybolt. US Secretary of Defense Robert McNamara who took office nine months later discovered that Skybolt was "a pile of junk", so it is no surprise that the UK government was back looking for a better option two years later. This time the merits of submarine-launched missiles over air-launched ones had been recognized, and the UK ended up persuading the US

---

[151] Mikhailov (1999) pp. 2.1, 2.3.

Figure 40. Britain's Black Arrow was canceled three months before it successfully placed the Prospero scientific satellite in orbit in October 1971.

to sell it the Polaris missile in return for allowing US bases to remain in Britain, including some stocked with nuclear weapons.[152]

Britain's missile program staggered on into the 1970s. Blue Streak became the first stage of Europe's ill-fated Europa rocket which never flew successfully—the challenge of mating a rocket stage designed in inches and pounds with one designed in meters and grams proved insurmountable. A separate purely British development built on the Black Knight test missile that had helped to refine the design of Blue Streak. Turned into the three-stage Black Arrow rocket (Figure 40), it eventually

---

[152] DeGroot (005) pp. 227–229.

launched the Prospero scientific satellite in 1971 making the UK the sixth nation to orbit its own satellite—after the Soviets, the US, France, Japan, and China. Three months before that successful October 28th 1971 launch, the government canceled all further work on Black Arrow. It had been a cheap and cheerful program costing less than $40 million, but you couldn't be in the space business in earnest on the cheap, and the UK finally realized this. Launching the 66 kg Prospero into a low orbit was intellectually interesting, but a long way from being able to put a satellite of a ton or more into geostationary orbit where it could do useful things. Black Arrow stretched its technology to the limit and there was no easy way to scale it up for more substantial space projects.[153]

Once the Europa rocket was canceled and its development agency, ELDO, merged into the newly created European Space Agency in 1975, British investment in satellite launchers came to an end. The only sign of this once successful industry was the Skylark sounding rocket, over 400 of which were launched until the late 1990s.

Given the American agreement through NATO to retaliate against any Soviet attack on Western Europe it is not clear how much of a real deterrent Britain possessed. If the USA were to abandon Europe then Britain's ability to destroy the dozen largest Soviet cities outside Moscow might carry some weight.[154] But since Britain relied on America for missiles, it seemed unrealistic that Britain would be able to continue as a serious nuclear power if it fell out with America. And how was a Soviet General to know whether the Polaris/Trident missile coming towards him came from a British or an American submarine? This last factor, alone, implies that the UK decision to launch a nuclear missile required US approval.

Many US leaders were not unhappy that Britain and France had a nuclear deterrent during the Cold War. The NATO agreements committed the US to defend Western Europe against Soviet attack, and this put America in the uncomfortable position as Henry Kissinger phrased it as having to "deter Soviet attack on Europe by risking Soviet destruction of America."[155] So, the French and UK nuclear deterrent was a potential way to avoid the US being dragged into a nuclear exchange with the Soviets. However, not everyone thought like this. Defense Secretary Robert McNamara was reported to be "worried that Europeans would use their nuclear weapons to trigger the thermonuclear holocaust"[156]—with good reason as we will see in the case of France below.

Britain participated actively with the US and the Soviet Union in negotiating the 1968 Nuclear Non-Proliferation Treaty, and the UK Government nowadays takes great pains to assert its compliance with that Treaty when discussing its nuclear arsenal. For example, to demonstrate its efforts to implement Article VI (to seek nuclear disarmament) the UK claims to be the only nuclear power to have reduced its delivery platforms to a single type—the Trident SLBM. It also highlights its

[153] Spufford (2003) pp. 6–36.
[154] Healey (1989) p. 455.
[155] Kissinger (1979) p. 220.
[156] Healey (1989) p. 244.

contribution of $750 million over 10 years to help dismantle Soviet nuclear and chemical warfare materiel.[157]

Britain's four nuclear-powered submarines ensure that at least one is on station at all times with a reasonable margin for accidents and occasional major refurbishment—a four-for-one strategy. While one is on patrol, another is preparing to take over, a third is in routine maintenance after being on patrol, and a fourth is assumed to be undergoing long-term refurbishment or otherwise not available. Each submarine can in principle carry 16 missiles each fitted with 12 MIRV bombs. However, each submarine currently carries no more than 48 warheads, with a varying and flexible mix of warheads per missile.[158]

Britain will have to decide in a few years' time whether to upgrade its submarines and their missiles. There are strong voices raised against the heavy expenditure either of those upgrades implies, but the present government has stated its intention to build a new class of submarines and to participate in the American program to extend the life of the Trident missiles. The somewhat luxurious four-for-one concept is now being re-considered as Britain prepares for a new round of investment. Britain has also committed to reducing its total number of warheads to 160 for its 50 Trident missiles.[159] We will return to the future of Britain's nuclear deterrent in Chapter 10.

Another feature of Britain's nuclear role in the world is the loss of capability and influence in the use of nuclear energy, thereby reducing its ability to implement Article 4 of the Non-Proliferation Treaty (help other countries use nuclear energy). In the 1950s and early 1960s Britain was perhaps the most advanced user of nuclear energy in the world, illustrated, for example, by its successful deployment of gas-cooled reactors. Since the 1980s no new nuclear power stations have been built, and even their low-carbon footprint has not persuaded the current otherwise enthusiastically green UK Government to begin a new building program. Nuclear now supplies less than 20% of UK electricity, down from 26% in the mid-1990s and destined to continue falling for at least another decade. Natural gas (40%) and coal (33%) are the main fuels used.

In the privatization frenzy of the Thatcher era, the electricity industry was sold to the private sector. Thankfully the regulatory mayhem inflicted on California and some other US States in their electricity privatization was avoided in the UK. In fact, the overall scheme was pretty sound, ensuring with a reasonable even-handedness that costs and benefits went hand in hand. One of the unforeseen consequences of being even-handed was that no private buyer was interested in taking on the nuclear power stations because of the unknown but enormous clean-up bill once the stations reached the end of their operating life and had to be decommissioned. The government spent a further 10 years working out an arrangement whereby the eight most modern reactors were privatized but at a price that was less than a fifth of their original construction cost. The less modern reactors (in which private buyers had no

---

[157] Secretary of State (2006) p. 13, 32–33.
[158] Barrie and Butler (2006).
[159] Secretary of State (2006) pp. 13, 26, and Barrie (2006).

interest), plus the facilities for recovering nuclear fuel from reactor waste and for disposing of that waste remained in public hands.

Where once Britain was able to help other countries exploit nuclear energy, and the relevant responsibilities under Article 4 of the Non-Proliferation Treaty, this is increasingly less the case. Government spending on research into nuclear energy is targeted almost exclusively at decommissioning problems and not at new developments. The once strong UK nuclear industry is now largely a part of the service sector, with undoubted capabilities in regulating and operating an efficient electricity generation and distribution business but less in designing and building new nuclear power stations.

One area of continued British know-how is in processing spent fuel—all that's left of a once ambitious plan to build breeder reactors (a reactor that creates more nuclear fuel than it consumes). But as with every other country, the challenge of safe long-term disposal or storage of radioactive waste remains unsolved.

A form of British dependence on the US linked to their history of nuclear cooperation is through the spy satellite imagery needed to manage Britain's nuclear arsenal. America provides the imagery and other intelligence to analyse an enemy's preparations and to locate targets under an arrangement that the British government used to call the "special relationship" with America but recently has been referred to as a "privileged relationship".[160] This trans-Atlantic connection dates back to World War II and the collaboration in strategic targeting during the bombing campaign in Germany. The two countries continued to collaborate on strategic reconnaissance into the Cold War, with the UK undertaking and hosting a number of high-risk over-flights of Soviet territory on behalf of the US. In the 1950s Britain began development of a surveillance satellite as the first passenger to be carried into orbit by the Blue Streak launcher, but once the launcher program was canceled the spy satellite developments also ceased. A technically ambitious ELINT satellite program called Zircon was canceled in 1987 when costs spiraled out of control and it was made public by the investigative journalist Duncan Campbell. Thus, Britain has no surveillance satellite of its own to offer to other countries to help defuse regional tensions—the kind of assistance the nuclear powers are supposed to provide to help achieve nuclear disarmament under Article 6 of the Nuclear Non-Proliferation Treaty. There are some signs of a change in policy on this matter which we will return to in Chapter 10.

FRANCE

The military logic for France's nuclear weapons was similar to that of Britain's—a small invulnerable nuclear force could deter attack even by a superpower because it

[160] Written evidence submitted to the House of Commons Select Committee on Science and Technology hearings on UK Space Policy—Submission from the Ministry of Defence, October 2006, p. 299.

Figure 41. President Kennedy (left) and French President Charles De Gaulle, 1961.

could inflict damage out of proportion to any benefit the superpower could hope to gain.[161] Unlike Britain, France was unwilling to consider depending on American weaponry, and so in addition to demonstrating an atomic bomb capability in 1960 and a H-bomb in 1968 it began construction of its own submarine fleet, the first one entering service in 1971, and its own surveillance satellites. France also refused to depend on European partners—an early initiative to cooperate with Italy and Germany was terminated by General de Gaulle as soon as he became President.

World War II was barely over when De Gaulle, as head of the Provisional Government, set up the Atomic Energy Agency—October 1945. However, communist influence was strong in post-war France and slowed the move to full-scale development. In 1952, the government finally made a real commitment to developing a bomb and ordered the building of a facility to produce plutonium. In December 1954 a decision was taken to actually build an atomic bomb itself, partly to try to rebuild France's international reputation after the French defeat in Vietnam by the Viet Minh rebels in April that year. De Gaulle returned to power in 1958 with renewed enthusiasm to create a French nuclear deterrent, and the first test of an atomic bomb took place in February 1960—at 60 kilotons the largest first test of any nation. De Gaulle saw the French bomb as a way to ensure that the US would come to the aid of Europe in the event of a Soviet attack—a way to *entangle* the US with Europe's destiny (Figure 41).[162]

Like Britain, France made sure to undertake atomic and hydrogen bomb tests well away from the homeland. The first four tests in 1960 and 1961 (all above ground and therefore creating radioactive fallout in the atmosphere) were conducted in the relatively flat Sahara desert region in southern Algeria—a French colony at the time. A further 13 underground tests took place in the more mountainous Ahaggar region of south Algeria until 1966. With Algeria about to be granted its freedom (and

[161] Healey (1989) p. 309.
[162] DeGroot (2005) pp. 233–235.

therefore not welcoming its former colonial master's nuclear explosions on its territory), and with plans to test a H-bomb imminent, France moved its test site to the Pacific. From 1966 until it stopped testing in 1996, France exploded 193 nuclear devices in the Mururoa and Fangataufa Atolls in French Polynesia. Forty-six of these were in the atmosphere (the last in 1974) provoking extensive international criticism. Even the underground tests in Polynesia were heavily criticized because of long-term damage to the coral reefs in the area. With the same cynicism as all of the nuclear powers, once it had completed the design of its warheads France halted its nuclear tests and began urging other countries to do likewise.

There is no sign of France giving up its nuclear deterrent as evidenced by its ongoing investment in an expensive upgrade of its submarine-launched missiles. The first successful test of the new M51 missile was completed in 2006. Shortly thereafter the Government placed a $350 million order for development of an upgraded warhead carrier and re-entry assembly that will be integrated with a new nuclear warhead also currently under development.[163] France has about 200 of the current M45 SLBMs, and these will be replaced in about 2010. The M45 has a range of 6,000 km and can carry six 100-kiloton MIRV bombs—the M51 is said to have a 50% longer range (which presumably could be traded for a heavier warhead).[164]

Like Britain, France has four nuclear missile-carrying submarines. It used to have a fleet of land-based ICBMs scattered across south and east France, and of classical aircraft bombs, but both of these have now been scrapped. It still maintains an airborne nuclear capability, using cruise missiles to deliver the bombs. A new version of the air-launched cruise missile has been under development for some time and is due to be fitted to the Mirage and Rafale fighter-bombers in 2008–09. The total French nuclear arsenal comprises about 350 warheads.[165]

France can point to several actions that respond to Article 6 (nuclear disarmament) of the Non-Proliferation Treaty. After the end of the Cold War, the gravity bomb (Hiroshima/Nagasaki form of delivery) and the land-based missile were scrapped, and one of the five missile-carrying submarines was withdrawn from service. Perhaps most significantly the production of weapon-grade plutonium at the Marcoule reprocessing plant and of highly enriched weapon-grade uranium at the Pierrelatte enrichment plant was halted in 1996, and since then France has produced no fissile material for nuclear weapons. The $7.5 billion funding to decommission the Marcoule plant has now been agreed, most of it from central government rather than the operating company. Plutonium extracted from spent fuel rods is turned into mixed-oxide (MOX) fuel, which can be used in many nuclear reactors across Europe. The process also produces reprocessed uranium most of which is stabilized for long-term storage, and the remainder enriched for re-use as reactor fuel (but this currently costs three times more than fresh uranium and causes operational difficulties).

[163] Taverna (2007).
[164] Dupont (2006).
[165] Secretary of State (2006) p. 15, and Dupont (2007).

The rationale for France's nuclear deterrent (the *force de frappe*) is not the subject of wide debate—whereas in Britain it is. French national pride seems to be a sufficient motivation for most of the French public and politicians, given that the nuclear force eliminates a dependence on either the US or Russia. When de Gaulle established the French Atomic Energy Commission in 1945 a major part of his rationale was the need to strengthen France's status as a major power which had been called into question by its rapid defeat by Germany in World War II. French diplomats now perceive themselves as leaders in any debate about European foreign policy, using their fully independent nuclear deterrent (unlike the British one that depends on US missiles) and its Permanent Seat on the UN Security Council to support that stance.

Having developed its own nuclear force, France also took the next obvious step and deployed its own spy satellites, something Britain has so far avoided doing. A recent statement by the French Defense Minister referred to enhanced efficiency of military operations as the rationale for the country's military satellites. The minister then went on to claim that these satellites also "enable the countries that possess them to assert their strategic influence on the international scene" and proceeded to put her money where her mouth was by proposing a 50% increase in France's military space budget.[166]

France has had a commercial surveillance satellite for over 25 years—the series of SPOT satellites. SPOT started out very similar to the US civilian Landsat series providing wide-area images in multi-spectral and stereo variants with a resolution of 10–30 m. The most recent SPOT-5 satellite in this series launched in 2002 has a resolution of $2\frac{1}{2}$ meters, and the next in the series to be called Pleiades will have 70 cm resolution. Although the stated purpose of the SPOT series has been commercial and scientific, its largest customers have generally been the military, and in particular the US military. The main military significance of SPOT images has been to improve maps for planning of military actions in poorly mapped regions of the earth. In the course of campaigns in Bosnia, Iraq, Afghanistan, Somalia, and other trouble spots the US military have found it convenient to use SPOT imagery to improve their maps. While US spy satellites could presumably provide a better quality of information, why divert them from high-priority tasks when the civilian imagery is perfectly adequate? Throughout the 1980s and most of the 1990s SPOT and Landsat had the market to themselves, but as we will see in Chapter 10 there are now several competitors offering similar imagery to all comers—for a fee.

The impetus for the expensive French investment into the military version of surveillance satellites (over and above the SPOT series) came after the 1990 Gulf War in which American satellites were the only source of high-resolution images, and not always available to French forces. With a small contribution from Italy and Spain, France launched the Helios satellite series starting in 1995, with the first of a second generation launched in late 2004 (Belgium and most recently Greece replaced Italy as partners in the second generation). The latest Helios satellites (Figure 42) collect visible images with resolutions better than a meter, and also have infra-red cameras giving some night-time capability. As one would expect with a modern satellite, and

---

[166] DICoD (2007) p. 3, and de Selding (2007a).

Figure 42. Helios-2A, France's second-generation optical military surveillance satellite.

as is the case with SPOT, the images are radioed to earth—avoiding the old-fashioned capsule return process.

Lacking radar, the Helios satellites are blind when clouds cover the scene. Rather than develop its own radar satellites, France has reached agreement with Germany and Italy to have access to the radar imagery to be produced by the spy satellites of those two nations under the German SAR-Lupe program (first satellite launched in late 2006, second in July 2007) and the Italian Cosmo-Skymed program (first satellite launched in June 2007). Belgium, Spain, and Greece have joined with these three larger European countries to study a collaborative next-generation system called Musis.[167]

[167] Lardier (2007c).

France is willing and able to provide imagery from its Helios satellites to friendly states. And it can also sell the satellites themselves, perhaps with performance slightly downgraded from its own variants. This capability is recognized in Third World countries as a useful alternative to relying on US or Soviet sources of information, and helps ensure that states that feel threatened by their neighbors can check the military stance of that neighbor reliably and without the need for provocative reconnaissance aircraft over-flights.

France has also become a world leader in nuclear energy. Almost 80% of France's own electricity is generated by nuclear power, the highest proportion of any of the nuclear states—and France generates more electricity than it uses, exporting 10–12% of the electricity generated to nearby countries especially Italy and more recently the UK. In fact, electricity is France's fourth largest export. The choice of nuclear over oil or gas for this purpose was taken after the oil crisis of the 1970s on strategic political grounds, not those of short-term economics (the British way). The 10,000 tons of uranium needed for these reactors each year is imported from a mix of stable and less stable countries: Canada, Australia, Niger, Kazakhstan, and Russia, providing a reassuring diversity of supply. In addition to insulating France from fluctuations in oil and gas prices, by reducing its dependence on importing those fuels, France feels more able to take an independent position in international debates that involve the oil-rich countries in the Middle East and the former Soviet Union. It is also able and willing to help other countries to build their own nuclear power stations —thereby actively implementing Article 4 (help other countries use nuclear energy) of the Non-Proliferation Treaty.

France's 59 nuclear power stations have so far avoided accidents like Three Mile Island in 1979 and Windscale in 1957 that undermined confidence in nuclear energy in the USA and Britain, respectively -the British ploy of changing Windscale's name to Sellafield did little to restore confidence. Even the Chernobyl disaster in the Ukraine did not dent the French commitment. France has a thriving research and development program in new types of reactor design, both to replace its own reactors and for export. It also has a successful export business in the extraction of fuel from reactor waste—its largest export earner with Japan, for example. Proof if it were needed of France's leading position in the commercial nuclear field came in 2004 when the US government entered into an arrangement to use the French *Phenix* fast-breeder reactor for research into new types of reactor fuel.[168]

CHINA—THE FUTURE SUPERPOWER?

The other nuclear power deeply engaged in Cold War politics is China. As told by Jung Chang and Jon Halliday in their ground-breaking biography of Mao Tse-tung, China negotiated the technology to develop nuclear weapons from the Soviet Union

[168] Uranium Information Centre (2007).

Figure 43. Mao Tse-tung (standing, centre) and Stalin (standing, left) watch Chinese Premier Chou En-lai sign a China–Soviet Union Treaty of Friendship, December 1949.

with a mix of guile, bluff, and above all sabre-rattling on a hair-raising scale.[169] Its first atomic bomb was tested in 1964 and the first H-bomb just $2\frac{1}{2}$ years later.

From the moment in 1948 when Mao's communist forces took power in China (Figure 43), apart from the island province of Taiwan (Formosa) 150 km off the southeast coast held by the remnants of Chiang Kai-shek's nationalist forces, Mao expressed to his colleagues a desire to possess nuclear weapons. In public his stance was the complete opposite; he called the Bomb a "paper tiger" which of course it had rapidly become—too destructive to use in anything but a world war situation—and professed his faith in the power of "the people" instead.

The history of the transfer of Soviet hi-tech know-how to China began in earnest with the Korean War in 1950. Egged on by Mao, Stalin allowed the Americans to be drawn into defending South Korea in its civil war with North Korea. Stalin instructed his UN Ambassador to stay away from the UN Security Council when it voted to send UN troops (largely American) to the aid of the South Koreans. If the Soviet Ambassador *had* attended that crucial Security Council meeting he could have vetoed the UN involvement in the war. Stalin's motivation was primarily to lessen American pressure in Europe, where the Soviets had tried to blockade Berlin into submission and failed because of Western airborne aid. Mao assured Stalin that he would pour millions of Chinese troops in to North Korea to engage the Americans in an expensive military stalemate—provided Stalin gave him the necessary jet aircraft, tanks, and other modern military equipment, a bargain Stalin was willing to accept. Mao's willingness to incur huge losses in the Korean conflict was partly because he

[169] Chang and Halliday (2005) pp. 425, 588.

assigned former Nationalist troops who had surrendered in 1948 to the battlefield, and felt no compunction about using them as cannon fodder to keep the conflict alive.[170]

Mao realized that by fighting the Americans he could negotiate military technology from the Soviet Union. As the Korean War dragged on he achieved one of his objectives when President Eisenhower in his inaugural State of the Union message on February 2nd 1953 suggested that he might use the atomic bomb against China. This veiled threat was enough for Mao to propose to Stalin that he give China the bomb so that the Soviet Union would not be drawn into a nuclear conflict with the US.[171] Eisenhower wrote in his diary that he hoped he had not made any blunders in this his first major policy speech, and based on the reactions he received he felt that he had not—a judgment that clearly didn't take account of Mao's reaction.[172]

Stalin inconveniently died a month later and Mao failed to extract nuclear secrets from his successors despite dragging out the end of the Korean War by three months—both the new leaders in the Kremlin and the new US President had decided to end the conflict.

Mao waited a year and then created another excuse to seek Soviet nuclear technology. In September 1954 he began a bombardment of the small island of Quemoy off the Chinese coast that was held by the Taiwanese Nationalists. The bombardment continued into the following year when Mao escalated the bombardment and built up forces along the coast as if about to invade Quemoy and its neighboring island of Matsu—seen as a prelude to a Chinese invasion of Taiwan itself. Mao also ratcheted up the tension by giving harsh prison sentences to 13 US airmen shot down over China during the Korean War, despite the fact that the armistice ending the war specified that all prisoners of war would be repatriated. Fending off hawkish advice to retaliate massively against China, Eisenhower responded by signing a mutual defense treaty with Taiwan, thus formalizing the protective role of US forces in the region. China increased the bombing of the off-shore islands, and made public statements that an invasion of Taiwan was imminent. The US fell for this ploy and in March Eisenhower and his Secretary of State John Foster Dulles both stated that small nuclear weapons would be used against China if the military situation demanded it.[173]

Mao now had what he wanted: a threat by the US to use nuclear weapons against China. Khrushchev had no desire to be dragged into a nuclear war with the US and made the historic decision to provide nuclear assistance to China—the Soviets agreed to build two of the key facilities needed for a nuclear bomb program: a cyclotron and a nuclear reactor, and agreed to train large groups of Chinese would-be nuclear scientists in the Soviet Union. A 12-year nuclear plan was drawn up with the help of Soviet scientists and other industrial assistance provided.[174]

---

[170] Chang and Halliday (2005) pp. 374–375.
[171] Chang and Halliday (2005) p. 390.
[172] Ambrose (1984) p. 49.
[173] Ambrose (1984) p. 239.
[174] Chang and Halliday (2005) pp. 414–415.

China still didn't have an actual nuclear bomb and Mao attributed his sidelining during the 1956 Suez crisis to that fact. His next opportunity to get nuclear technology from the Soviets came in 1957 when Khrushchev needed his support at the big Communist Summit to be held in November in Moscow. Mao had been critical of the Soviet leaders since Stalin's death and Khrushchev judged that he needed to sweeten the Chinese leader in order to have his full support at the November Summit. In October, therefore, the Soviet Union agreed to hand over an atomic bomb to the Chinese together with experts and materiel to enable China to construct its own. Just as Soviet clandestine access to the American and British research saved them years in developing the bomb, so too did the Soviet assistance to the Chinese. Nevertheless, the price tag of the development was enormous (Chang and Halliday quote a US estimate of $4.1 billion in 1957 prices) at a time when China was barely able to feed its inhabitants—and the cost of the nuclear development probably contributed to the massive 1958–1961 famine that took 38 million lives. The drain on China's resources continued for many years causing power shortages, for example, as the electricity-hungry nuclear processing facilities began production.[175]

Although Khrushchev withdrew assistance with the bomb in June 1959 over Mao's refusal to hand over the guidance system of an undamaged American air-to-air Sidewinder missile that dropped from a Taiwanese plane in September 1958, China had enough knowledge to proceed. The first test of the Chinese bomb came on October 16th 1964, and as recounted in Chapter 5, the device was a 22-kiloton uranium bomb and not the plutonium bomb the US was expecting. The Chinese announcement of the test heralded it as a totally Chinese development achieved without any foreign assistance, which gave rise to great pride among the population.

Following the first atom bomb test, China continued development towards an H-bomb. Two tests in 1966 were demonstrations of the principles of a fusion device, the second of these being a demonstration of the two-stage Teller–Ulam principle (Figure 11). Six months later on June 17th 1967 came China's first H-bomb test, with a yield of 3 megatons—only $2\frac{1}{2}$ years after the explosion of its first atomic bomb. Only one other test (in 1976) had a larger yield, reaching 4 megatons.[176]

Chinese tests have all taken place at the Lop Nur test site on the edge of the Tarim Basin in the northwest of the country—47 tests in all, of which 23 were above ground. Between 1980 and the last test in 1996 all were underground.

Mao's next requirement in his quest to join the superpower club was a submarine-launched missile system, and he found an excuse to press the Soviets for this before the Sidewinder affair poisoned the relationship. He repeated the Taiwan ploy used four years earlier and in August 1958 started a new barrage of Quemoy. America naturally thought Mao was building up for an invasion of Taiwan—he had threatened as much many times before. Eisenhower was worried about "devious Orientals" but he was referring to the Nationalist leader Chiang Kai-shek (who wanted to drag the US into a war so that he could win back the Chinese mainland) not the even more devious Mao. In his memoirs Eisenhower

[175] Chang and Halliday (2005) pp. 425–426.
[176] Mikhailov (1999).

blamed the Soviets for initiating the crisis, illustrating how little the West understood the politics between the two big communist powers. At the time, Eisenhower rebuffed various suggestions from the Secretary of State and from the Joint Chiefs that he use nuclear weapons to make the communists back down, but he did authorize a strengthened Seventh Fleet to assist the Taiwanese short of attacking the mainland.[177]

Khrushchev wanted no part of a Taiwanese war, potentially a nuclear one, and he bought Mao's acquiescence with a deal to provide submarines and missiles to fire from them. Throughout late-1958 the Soviets agreed to transfer various defence technologies to China culminating in February 1959 with an agreement to supply ships and weapons including missile-firing submarines and the missiles to fire from them.[178]

China's current strategic deterrent is said to comprise a silo-based ICBM force of around 20 missiles. It also deploys a larger number of nuclear-armed intermediate- and medium-range ballistic missiles, all of which are believed to carry single warheads. New projects include mobile ICBMs, an ICBM equipped with multiple warheads, a submarine-launched strategic ballistic missile, and (potentially nuclear-capable) cruise missiles.[179]

The US considered bombing the Chinese nuclear facilities in the early 1960s—as the Israeli's did to the Iraqi nuclear facilities 20 years later (see Chapter 9). In July 1963 President Kennedy instructed Ambassador Harriman to seek Khrushchev's agreement for such a strike. Khrushchev rejected the idea—it was his technology after all. Kennedy continued to consider his options for an attack—the plutonium plant at Baotou 500 km west of Beijing and the gaseous diffusion plant at Lanzhou, a further 500 km southwest of that, were two of the targets considered for a bombing raid. The possibility of Taiwanese commandos sabotaging the Lop Nur test site was also discussed—this seems pretty far-fetched since Lop Nur in the far west of China is as distant from Taiwan as is Australia. None of these plans came to anything, but Mao was aware of the American threat and made it known that he would throw Chinese troops into the Vietnam conflict if any such US attack occurred, bogging the Americans down there even more than they were already.[180]

The thaw in US–China relations during the Nixon Administration resulted in China receiving some Western nuclear technology. In addition to the agreements for US listening stations in western China described in Chapter 5, Kissinger and Nixon agreed to allow China to buy French and British technology for commercial nuclear reactors even though the export of such sensitive technology was prohibited.[181]

Despite a backward economy and the setbacks of the Cultural Revolution, the 1959 missile technology agreement with the Soviet Union enabled China to launch its first satellite on April 24th 1970—recent television interviews with participants in the

---

[177] Ambrose (1984) pp. 482–485.
[178] Chang and Halliday (2005) pp. 430–432.
[179] Secretary of State (2006) pp. 15–16.
[180] Chang and Halliday (2005) pp. 500–502.
[181] Chang and Halliday (2005) p. 613.

satellite program have confirmed that the Cultural Revolution caused many months' delay in the launch. That delay was particularly significant and made the Chinese achievement somewhat anti-climactic at the time because the launch came just $2\frac{1}{2}$ months after China's historical east Asia rival, Japan, had launched its first satellite. China's launch of humans into space in 2003 can be seen as a statement that China has finally overtaken Japan in the space business—only America and Russia having previously achieved this feat. In addition to Soviet technical assistance with its rocket program, the Chinese also benefited from some US technology that they bought from Fidel Castro when a Thor rocket fell on Cuba—Castro had the Soviets and the Chinese bid against each other for the debris and the Chinese got some technology that apparently assisted their developments significantly.[182]

Development of a reliable ICBM took longer, with the first successful test flight coming in 1980.[183]

China now has a long-established range of spy satellites, starting with capsule return techniques and in recent years moving on to digital camera technology and radio links. The first capsule return satellite weighing a massive $1\frac{3}{4}$ tons was launched in 1975 making China only the third nation to achieve the feat of recovering a capsule from orbit. The return capsule was recovered after a three-day flight.

China never attempted to match the rapid launch rate of these satellites that America and the Soviet Union had practiced during the 1960s. The second Chinese mission came a year after the first one, and the third a year after that. The launch rate remained at just less than one per year through the 1990s, and with a very high success rate—only one of the first fifteen flights failed to return its capsule successfully. The stay in orbit before returning the capsule gradually increased first to five days in 1982 and then to eight in 1987 presumably reflecting a larger film capacity.

With an increase in their size to 2.1 tons in 1987, and then to 2.6 tons in 1992, China began to double up the objectives of these satellites. From 1987 the satellites and return capsules carried so-called microgravity experiments—these are scientific experiments designed to explore the consequences of the almost zero gravity (hence *micro*-gravity) found in orbit. The Chinese explored problems in physical chemistry such as crystal growth, and in biology such as algae growth. On the 1990 flight they carried two guinea pigs and successfully returned them—another example of China being the third nation to achieve a space milestone, this time flying animals in orbit.

In the 21st century China has taken a number of dramatic initiatives in space— the launch of humans in 2003 and 2005, and then the controversial attempts to blind US spy satellites using lasers in 2006 and the destruction of one of its own obsolete satellites with an anti-satellite missile in January 2007. We will discuss the implications of these developments in Chapter 10.

---

[182] Chang and Halliday (2005) p. 488.
[183] Chang and Halliday (2005) p. 588.

9

After the Cold War—regional tensions

Nine countries are now considered to be nuclear powers, with North Korea as the most recent member of the club and Iran threatening to bring the number up to 10. The others are the USA, Russia, France, Britain, China, Israel (allegedly), India, and Pakistan.

Twelve countries possess surveillance satellites with a resolution that makes them equivalent to satellites like CORONA from the 1960s, and so would have been considered to be spy satellites in that era. The 12 include 7 of the nuclear powers—Pakistan and North Korea being the two nuclear powers without such satellites. The non-nuclear countries in this group are Taiwan, Japan, South Korea, Italy, and Germany. The list keeps growing with satellites of this type under construction for the United Arab Emirates and Algeria, to name but two. Most of these satellites are commercial or scientific and have resolutions ranging from 2.5 m down to 50 cm. The proliferation of these satellites started in the mid-1990s when Russia began to sell imagery from its military satellites on the open market, and the US permitted two American companies to build and operate purely commercial satellites offering images with resolution better than 1 m.

Surveillance satellites that are specifically military in nature are operated by seven countries, namely Russia, the US, France, Japan, Germany, China, and Britain (although Britain's is only a technology demonstrator satellite, not an operational system). The impetus for new countries to build these satellites has come from the fragmentation of the world's military threats since the end of the Cold War. Where before the main threat was a US–Soviet confrontation—either directly or via satellite states—military forces from the developed countries are now involved in actions across the globe. The proliferation of missile and nuclear technology has also motivated countries to have an autonomous satellite-monitoring capability. Japan's decision to build a fleet of radar and visible imaging satellites stems from concerns about the missile tests undertaken by North Korea.

As concerns strategic military affairs and spy satellites, Russia has taken over the role formerly played by the Soviet Union. Since the end of the Cold War in about 1989 both Russia and America have agreed various major reductions in their nuclear arsenals—speeding up the process begun hesitantly with SALT-I and SALT-II.

Let's run through these various countries in turn.

SOVIET UNION AND THE USA

With the coming to power of Mikhail Gorbachev in the Soviet Union in 1985 arms control negotiations progressed rapidly. At that time, the US had over 11,000 strategic warheads and the Soviets 10,000. If you add in the warheads on intermediate-range and short-range missiles the figures become 21,000 (US) and 20,000 (Soviet)—an absurd level of over-kill.[184] The Intermediate-Range Nuclear Forces (INF) Treaty was signed in December 1987 addressing the huge numbers of shorter range missiles targeted at and from Europe. This was followed by the first Strategic Arms Reduction Treaty (START-I) in 1991, which cut the numbers of long-range nuclear weapons by roughly half. Then two years later START-II agreed to scrap a further 3,000–3,500 warheads.[185] Most recently, the Treaty of Moscow in 2002, signed by Presidents George W. Bush and Vladimir Putin (Figure 44), reduced the number of weapons by a further 1,700–2,200.

The need for spy satellites has not however reduced. Monitoring the exact make-up of the other superpower's missile fleet is no longer quite so difficult because these more recent Treaties include on-site inspection clauses. The satellites are increasingly used in support of tactical military operations rather than the longer term strategic objectives of Treaty verification. Of course, expensive satellites like Big Bird were always called into service to support major military operations—for example, in the first Gulf War in 1991. There has inevitably been a tension between the front line military services that want information about the here and now, and the intelligence services that want to build up a total picture of an enemy with a view to forecasting his intentions. For now the compromise continues whereby satellites are tasked to do either strategic or tactical imaging depending on changing priorities. The limitations of this state of affairs became apparent during the first Gulf War. US satellites detected the build-up of Saddam Hussein's troops on the Iraq–Kuwait border, but the intelligence community was unable to advise on his intentions.[186] Later, when US troops were involved during the actual war, lots of imagery was made available to the forces near the front line, but they had not been trained how to interpret them. After Saddam's forces had been evicted from Kuwait, much time and effort went into improving the flow of information from satellite to analyst to local force commander, so that US forces engaged in the second Gulf War had significantly better information.

[184] DeGroot (2005) pp. 308–309.
[185] Blair *et al.* (1997).
[186] Lindgren (2000) pp. 184–185.

Figure 44. May 24th 2002, in the Kremlin in Moscow President Putin (left) and President George W. Bush sign the Treaty of Moscow that stipulates further reductions in Soviet and US strategic weapons.

The US has attempted to create a new spy satellite design that specifically addresses tactical military needs. Such a system would provide continuous or nearly continuous imaging of critical parts of the world, thereby preventing an enemy from concealing his activities as is possible at the moment from the occasional and predictable over-flight of today's spy satellites. It would also detect moving targets such as mobile missile launchers that have proved elusive in recent wars. Finally, it would include high-resolution radar imaging to provide 24-hour-a-day 365-days-a-year coverage. This grand plan has so far proved too expensive to implement, and the US is currently re-evaluating what it can do at an affordable price. In the meantime while this great plan slowly and unsteadily unfolds, during the 1980s and early 1990s there was usually one or at most two of the advanced versions of the KH-11 Big Bird introduced in Chapter 4 in orbit, each new one no doubt incorporating incremental improvements over earlier versions. Now in 2007 with US forces engaged in active combat in two or three places around the globe, there are four in operation launched in 1995, 1996, 2001, and 2005.

Since 1988 the US has also had a radar-imaging satellite in orbit, usually called by the name Lacrosse in the press—the military don't even admit to Lacrosse's existence let alone confirm its correct name. These satellites have the great advantage that they provide imagery day and night and in all weathers. But it has taken a long time for such satellites to appear.

In 1978 NASA launched the first civilian radar-imaging satellite, Seasat. It used an imaging technique called synthetic aperture radar (SAR[187]) to create an image out

---

[187] Another form of satellite also confusingly uses the acronym SAR: Search & Rescue, and is totally different from synthetic aperture radar.

of radar echoes, and had the enormous advantage of working day and night, and under all weather conditions.

Radar had been known to be much preferable to optical or infrared imaging as concerns its all weather day and night operation, but such satellites are inherently more expensive and complex. A spy satellite with a camera that takes optical pictures is passive, using sunlight to illuminate the scene. A radar satellite has to transmit radio waves to illuminate the scene and then detect the very faint echoes. The radar satellite's targets are more than a hundred kilometers away, so its transmissions have to be strong and its receivers (that pick up the echoes) have to be very sensitive. A radar satellite therefore needs lots of electrical power with which to drive its transmitters, which makes it big (i.e., expensive).

Turning radar echoes into an image requires complex electronics, which could not realistically be placed on a satellite until the 1970s. The first aerial imaging radars that used the SAR technique were extremely bulky, incorporating a mix of optical and electronic processing in order to achieve the necessary throughput. It wasn't until the performance of electronics improved to the point where they could handle the real-time part of the processing in an airborne SAR that the space-borne version was seriously considered.

As its name implies, Seasat was intended to study the oceans—its radar used a radio frequency that was particularly sensitive to water. It detected the height, direction, and roughness of waves across the world no matter what the weather—the first time such statistics could be collected on that scale. Seasat also detected ships and anything else in the water—and it detected them very clearly, again irrespective of weather or day/night conditions. Then after just 105 days in orbit Seasat failed due to a short circuit in the satellite's electrical system. It had taken just 42 hours of imagery.

One of the things Seasat could detect clearly was the wake of anything moving in the water. Rumors circulated shortly after the failure that it was too good at detecting submarines, even when submerged—presumably in the form of the wake left by the conning tower or the periscope or by the swell induced by their movement below the surface. Radio commands to fix the short circuit were sent too late from the NASA station in Chile so that by the time the satellite appeared over the next station it was dead. Needless to say the conspiracy theorists reckoned that the "short circuit" was induced by the American military in order to prevent the embarrassingly high-quality data being made public.

The US military had launched a radar satellite in 1964 called QUILL. The technology of the day meant that it had to record the images on a magnetic tape then return that tape to the ground in a recovery capsule, just like the CORONA satellites. The resolution of the images was not adequate for intelligence purposes, so the program was canceled.

So, it was 10 years after Seasat that America launched spy satellites carrying modern SAR imagers—the Lacrosse series. Initially, only one of these expensive 15-ton satellites was in orbit at any given time—a price tag of $1 billion per satellite is said to apply to each Lacrosse as well as to the advanced KH-11 Big Bird—and a further $500 million to launch each of them. Since 9/11 and the ensuing wars in

Afghanistan and Iraq there are often four Lacrosse satellites in orbit; the current batch of four Lacrosses were launched in 1997, 1999, 2000, and 2005, which means that a replacement for the longest serving one will be needed before long.[188] They are extremely large (50 metres across with their solar arrays deployed) in order to generate the 10–20 kW of power needed to drive the SAR—and correspondingly expensive. Having four Lacrosse and four advanced KH-11s in orbit together means that imagery of, say, Iraq or Iran is obtained every few hours, making it difficult to hide from them.

In addition to these eight wholly military satellites the US has access to imagery with resolution of panchromatic images (what you and I would call black and white) better than 1 m from the various commercial high-resolution systems now available: GeoEye's Ikonos-2 (launched in 1999 and offering 1 m resolution) and soon GeoEye-1 (planned for 2007, 40 cm); Digital Globe's Quickbird-2 (2001, 60 cm) and Worldview-1 (2007, 50 cm); and ImageSat's (Israel) Eros-B (2006, 70 cm). These satellites also provide multi-spectral images with somewhat poorer resolution, typically 2–5 m. During the 2001–2002 war in Afghanistan, the US bought up all imagery of that region from the then in-orbit satellites in this category, and has recently awarded $500 million contracts for five years' supply of imagery to GeoEye and Digital Globe.

Russia too maintains pretty much the same spy satellite designs that it had during the Cold War (see Chapter 4) and like the US engages in a mix of totally military and commercial systems. The multi-reentry capsule Yantar-4 satellite series is still in use under various designations. The Neman variant (Yantar-4KS1) transmits its images to ground either directly or via the Potok geostationary satellite, and each satellite has a life of about 1 year. For close-up imagery Russia still relies on the capsule return variant Kobalt (Yantar-4K2). Recent variants have had a lifetime extended from 2 months to about 4 months suggesting an increase in the number of capsules from the original two plus the camera itself. The massive 12-ton Orlets satellites are said to carry 20 return capsules and stay in orbit for a year or so.[189] A new type of electro-optical spy satellite called Arkon was first launched in 1997. Its orbit is unusual in that it comes no closer than about 2,000 km to the earth, but its apogee or highest point is even more unusual at 30,000 km. One website gives its resolution as 2–10 m and mentions its ability to loiter over a target for several minutes presumably in the sense that the camera stays pointing at a target while the satellite moves along its trajectory—a feature that its higher-than-usual orbit makes feasible.

Some of the Arkon images are thought likely to become available commercially in due course. But Russia already has a commercial series of surveillance satellites—the Resurs (sometimes called Ressource) series of which the most recent is the Resurs-DK launched in June 2006 and offering images with a resolution of 1 m (or perhaps even 80 cm) panchromatic and 2–3 m multi-spectral. This $6\frac{1}{2}$-ton satellite is based on the military Yantar satellite body. It is much heavier than equivalent Western satellites—the Israeli Eros-B, for example, weighs only 300 kg by

[188] *Air & Cosmos* (2006), and Covault (2006a).
[189] Clark (2001).

comparison. The prototype for a new class of surveillance satellite called Monitor was launched in 2005. Weighing only 750 kg Monitor is intended to provide imagery in the 1 m class much more affordably than the older generation of satellites.

Russia seems not to have caught up with the West in terms of imaging radar satellites. A hugely expensive system called Almaz-T was designed in the late 1970s. Like the Hubble Space Telescope it was designed to be visited by astronauts to be refurbished from time to time. Two of these giants actually flew from 1987 to 1989 and from 1991 to 1992 but without the man-tended features. Since then the manufacturer of the Almaz has been proposing a commercial variant on a much smaller satellite—advertising in Western trade magazines on a regular basis—but no sales have been reported.

The general public became aware of Soviet radar satellites when Kosmos 954 crashed in Canada's Northwestern Territory in January 1978 spreading radioactive debris over a wide area. The Soviets had decided to produce electrical power for the satellite using a radioactive generator in addition to solar cells. A radioactive generator has the advantage of working in the absence of sunlight—which low-orbiting satellites experience for as much as a third of the time. Most satellites get around the absence of sunlight by carrying large batteries that are charged up when the sun is visible and discharge in darkness. The disadvantage of using radioactive material to produce electricity is that if the satellite crashes you create radioactive debris—this awkward fact had persuaded Western nations to avoid the technology except for space probes heading far out in the solar system where sunlight is too weak, and even these missions provoke public demonstrations because of the risk of a crash (e.g., if the launcher explodes). As mentioned before, radar satellites need lots of electrical power, so the Soviets decided to take the risk of a crash.

The Soviets had to compensate Canada to the tune of several million dollars for the cost of the Kosmos clean-up, but this didn't stop them using the same radioactive technology on later satellites.

Note that Kosmos 954 and similar satellites that went under the Western title of RORSAT, used radar in the traditional sense, not for imaging. Their main objective was to pick up the clear radar echoes produced by shipping. The blip on a circular screen produced by an aircraft detected by a ground-based radar gives an idea of the information these satellites provided. This form of radar requires much simpler electronics than the imaging kind on the Lacrosse satellites. The US too has radar satellites for monitoring shipping about which little has been published.

INDIA AND PAKISTAN

Having exploded an atom bomb in 1974, India conducted no further tests until 1998. It got the plutonium for its first bomb from a reactor supplied by Canada containing heavy water supplied by the US. Its motivation seems originally to have been to respond to China's aggressive moves in the 1960s culminating in an invasion across India's northern border in October 1962. The conflict ended in a stand-off (with parts of the border still disputed to this day), but India felt let down by both America and

Figure 45. Launched May 5th 2005, India's Cartosat-1 satellite provides imagery with 2.5 m resolution and has stereo capability; the smaller Cartosat-2 launched on January 10th 2007 has a single camera with resolution better than 1 m. Credit: ISRO.

the Soviet Union—Mao Tse-tung had deliberately chosen the moment when the superpowers were engaged over Cuba, and had pre-arranged with Khrushchev that planned sales of MiG-21 fighter planes to India would be embargoed.[190] By the 1990s, however, the threat seemed more real on India's western borders– with Pakistan— where full-scale fighting had broken out three times since the splitting of the Indian sub-continent into three countries (Bangladesh being the third) in 1947.

India had also invested heavily in satellite technology and had launched its first satellite in 1975. Starting in 1988 it has orbited a series of IRS (Indian Remote Sensing) satellites ostensibly for civilian applications like agricultural management and flood monitoring, but with obvious military uses too. These satellites have had a resolution in the 5–30 m class and so were capable of monitoring the disposition of armed forces in Pakistan, China, and Myanmar, with all of whom India has from time to time had a difficult relationship, but they were not suitable for detailed intelligence. Since 2001, however, the TES and Cartosat (Figure 45) satellites have been acknowledged as having both civil and military applications and provide images with a resolution of about 1 m.

India has created an impressive space capability without vast funding or detailed foreign assistance. It has a range of launch vehicles able to place surveillance satellites into relatively low orbit, and communications satellites into geostationary orbit. The same technology of course gives it a considerable capability in long-range military missiles. The Prithvi has a range of 150 km and the newer Agni-3 3,000 km. According to a recent UK government report India can deliver its nuclear weapons via land-based missiles and is developing a submarine-launched ballistic

---

[190] Chang and Halliday (2005) pp. 487–488.

Figure 46. A. Q. Khan, "father" of Pakistan's atomic bomb program, acquired nuclear technology from Europe and later sold nuclear secrets to North Korea, Libya, and Iran.

missile capability. It also has versions of its nuclear bomb for delivery by bomber aircraft and is working on cruise missiles.[191]

India's deterrent and its spy satellite capability are now intrinsically linked to its relations with its neighbor Pakistan—and even more so in the reverse direction.

The story of how Pakistan obtained atomic technology is now well known.[192] Young engineer A. Q. Khan (Figure 46), filled with zeal to revenge his country's humiliation in its conflicts with India, emigrated to work in the Netherlands and got hold of the blueprints for uranium enrichment technology from his employer—the employer's security was pretty lax it seems. Returning to Pakistan he was given unlimited funding and gradually built up the infrastructure to manufacture weapons-grade fuel, and to manufacture the weapons themselves, buying further knowledge and technology from China and commercially from the West. His efforts culminated in the May 1998 underground tests of five nuclear devices—because the bombs were wired to fire simultaneously the exact number of devices is difficult to verify and is suspiciously similar to the number of Indian tests two weeks before (see below). Pakistan then exploded another bomb two days later to "go one up in the series" with India. The story of his activities might not have been revealed had he not begun to sell the technology to Libya, North Korea, and Iran, allegedly at least in part for personal financial gain, whereupon the story gradually leaked to the intelligence services and police forces in Western countries.

Just two weeks before the Pakistani tests, India conducted five underground tests—the first since its inaugural test in 1974, of which at least one device was apparently a fusion bomb, albeit with a yield well below 100 kilotons. The tests were perceived with all the more alarm in Pakistan because the test site was a bare 150 km from the Pakistani border. A month before the Indian tests, Pakistan tested its own long-range nuclear missile, the Hatf-5 (or Ghauri), which has a range of up to

[191] Secretary of State (2006) p. 16.
[192] Langewiesche (2005).

1,500 km, thereby making its contribution to boosting the tension between the two countries.

The India–Pakistan relationship is perceived by many to be the most likely place for a nuclear war to occur. At the moment each side's nuclear arsenal acts as a deterrent. But one worrying scenario is if Pakistan should come under the sway of an extremist Muslim regime, unlike its current relatively secular government. If the logic that impels suicide bombers in Iraq and elsewhere (just think of 9/11 in the US and 7/7 in London) were to prevail in the corridors of power in any nuclear state, who knows what actions such a regime might take? And Pakistan has the Ghauri missile whose range makes it a potential threat across the whole south and west Asia region.[193]

Unlike India, Pakistan lacks its own surveillance satellites, although it has launched two small prototype surveillance satellites—the most recent in 2001. It obtains imagery from commercial sources, which in recent years has included images with resolutions of better than 1 m. These sources of imagery are adequate in peacetime, but it remains to be seen if—in time of tension with India—Pakistan will feel compelled to orbit a spy satellite of its own. Its fledgling space agency focuses on building up the capability to process imagery from the satellites of other countries, which as we have seen above there are now in abundance. The current close ties of the Pakistan Administration to the US in connection with the war in Afghanistan presumably gives the Pakistani military access to American spy satellite imagery on request—with a fair chance of the request being accepted.

ISRAEL

Since its foundation as a state in 1948, Israel has been under constant threat from one or other of its neighbors—understandably since the 1917 UK Balfour Declaration that set the process of Israel's statehood in motion over-rode the *de facto* rights of the existing inhabitants of the region. The nuclear deterrent that Israel has developed plus its missiles and satellites are a response to this environment of almost continuous threat of war. The official Israeli position about its bomb is that it "has no comment". The policy is one of deliberate ambiguity: neither confirming nor denying that Israel has nuclear weapons—and official spokesmen are quite open about this, stating that they are being deliberately ambiguous on the subject. The policy is intended to create doubt in the minds of Israel's potential adversaries as to Israel's ability and willingness to use a nuclear bomb. Like other nuclear powers Israel has not found a use for the bomb in its several recent conflicts. It is said to have had two bombs armed and ready during the June 1967 Six-Day War and a dozen or so during the October 1973 Yom Kippur War, but this story may have been concocted as part of a policy of deliberate misinformation. Since Israel is one of the few countries that has not signed the Nuclear Non-Proliferation Treaty, the inspectors from the IAEA are not able to

---

[193] Richelson (1999) p. 228.

check what's going on at the vast Dimona complex in the Negev desert which is apparently where Israel makes its bombs.

The Israeli bomb first became public knowledge in 1986 when an Israeli nuclear technician, Mordechai Vanunu, sold information on the Dimona plant to the London *Sunday Times*. Vanunu described an underground facility where bombs and bomb materials were manufactured, including lithium deuteride required for an H-bomb. In a classic "honey trap" spy ploy, Vanunu was lured to Rome by a glamorous Mossad agent, where he was drugged and smuggled back to Israel to be tried and imprisoned for 18 years—of which 11 were in solitary confinement to prevent him discussing his nuclear secrets with other inmates. Released from prison in 2004 having served the full 18 years, legal injunctions mean that he is severely restricted in his movements and what he can say in public. Estimates of the size of Israel's nuclear arsenal range from 75 to 400 bombs. No Israeli nuclear test has ever been detected, with the possible exception of the 1979 "event" off the coast of South Africa described in Chapter 5.

Another response to the threats from its neighbors has been to develop a series of surveillance satellites. The Offeq series is for strictly military use. The latest, Offeq-7, is acknowledged as having a surveillance mission and was launched in June 2007 on the Shavit launcher which is a derivative of the Jericho-2 ballistic missile. The Eros series is operated by ImageSat on a commercial basis; Eros-B, launched in 2006 on a Russian launcher, provides images with a resolution of about 70 cm.[194] Weighing only 290 kg (compare the Soviet's 12-ton Orlets and the US's 13-ton Big Bird), Eros-B is an example of a new generation of surveillance satellites (Offeq-7 is similar) that exploit clever optics design and advanced detectors to provide very high resolution in a small and relatively cheap package. Also weighing about 300 kg, TechSAR is to be launched in 2007–2008 and will be Israel's first SAR satellite.

Israel also has an impressive missile capability. Its Jericho-2 has a range of at least 1,500 km which puts all of Iraq, Syria, and Egypt within range and large parts of Saudi Arabia and Iran—but the Iranian capital of Tehran is about 2,000 km distant and the Pakistani capital of Islamabad about 4,000 km. Even more impressive are the Israeli developments in anti-missile technology. With the help of US funding, the Arrow missile has become a viable candidate for an anti-missile defensive shield in the near future. Such a defensive weapon would provide some protection against the threat posed by Iran's ballistic missiles—the Ghadr-110 is said to have a range of over 2,000 km,[195] more than enough to cover the 1,200 km from western Iran to Jerusalem.

Having been the target of Scud medium-range missiles from Iraq in 1991 and of short-range missiles from Lebanon and the Gaza Strip on frequent occasions, Israel is aggressively pursuing an ambitious deployment of a defense against both types of missile. The threat from short-range missiles was emphasized during the summer 2006 war with Hezbollah in southern Lebanon. Various Israeli and US technologies are being considered for the shield against short-range missiles including high-power

[194] Covault (2006b).
[195] Covault (2007b).

lasers, commercial lasers, rocket interceptors, and submunition cannon (a sort of sophisticated cross between a shot gun and a machine gun), with the rocket interceptor being the front runner.[196] The Israeli Arrow system is likely to be the basis for the defense against longer range ballistic missiles.

Israel has shown itself willing to take huge risks to prevent its Arab neighbors acquiring a nuclear capability. The most dramatic illustration of this came in 1981 when Israeli planes bombed and destroyed the Iraqi nuclear facilities at Osiraq near Baghdad to prevent Iraq constructing an atomic bomb. Iraq of course denied that it was diverting plutonium from the French-built reactor to a bomb program, but even IAEA confirmation that no such diversion was occurring failed to convince the Israelis and they sent in the F-15s and F-16s from Etzion Air Force base 1,100 km away, flying low over Jordan and Saudi Arabia to reach their target. In the light of Israel's willingness to initiate preventive aggression against embryonic nuclear facilities in potential adversarial countries (primarily Islamic), Iran has taken care to distribute its nuclear facilities around the country, and Pakistan installed heavy air defense units around its nuclear facilities.

NORTH KOREA

After many threats to do so, North Korea finally exploded a plutonium bomb on October 9th 2006 near P'unggye. It was a bit of a damp squib, far below the 4-kiloton explosion North Korean officials had warned the Chinese to expect. Nevertheless, although "only" half a kiloton in terms of yield, the device has been confirmed as a plutonium implosion fission bomb based on sampling of airborne debris that is given off even from underground tests, and on seismic signals received at the Mudanjiang seismic station in China 370 km north of the test site.[197]

The purpose of the nuclear bomb is probably at least as much economic as military. The US is deterred from invading North Korea by the knowledge that any incursion could result in devastating North Korean damage using non-nuclear weapons to the South Korean capital, Seoul, a mere 40 km south of the border. Possession of a nuclear weapon adds little to that threat. However, possession of the ability to make nuclear weapons is a powerful bargaining chip in negotiations with America, South Korea, and Japan on any aspect of trade, economic aid, or other forms of economic interaction. North Korea discovered that it was ignored by the US during the late 1990s, with the Clinton and Bush Administrations showing little enthusiasm to come to the negotiating table. The threat, now a reality, of creating a nuclear arsenal seems to have brought the US back to the table.

North Korea has possessed medium-range missiles for more than a decade, the Taepo Dong 1 with a range in excess of 2,000 km. The much larger Taepo Dong 2 missile is under development—there was at least one failed test in July 2006. It is expected to have a range well in excess of 10,000 km, probably capable of reaching

---

[196] Opall-Rome (2007).
[197] Collins (2007), and Simpson (2007).

Europe or America.[198] Another feature of North Korea's missile activities is a willingness to make its know-how available to friendly nations such as Iraq, Iran, and Libya. The No-dong enhanced variant of the Scud is sold on the export market and has a range of 1,300 km.

The 1998 launch of its own small satellite used the Taepo Dong 1 launcher with a third stage on top, demonstrating—if that were needed—the sophistication of North Korea's missile technology. Foreign observers expressed doubt that the satellite actually entered orbit, but even if there was a fault with the upper stage the event showed that North Korea has the basic technology for placing satellites in orbit.

THE REST OF EAST ASIA—JAPAN, TAIWAN, AND SOUTH KOREA

As a non-nuclear power, Japan had relied on US strategic facilities until the North Koreans began testing missiles in the mid-1990s that could span the Sea of Japan separating the two countries—600 km at the nearest point. With considerable capability in scientific and commercial satellites built up over the preceding 20 years Japan began a crash program to deploy both optical and radar military surveillance satellites. The first Information Gathering Satellite (IGS) was launched in 2003, and deployment of the constellation of two optical and two radar satellites was completed in February 2007, although a month later the older of the optical satellites experienced severe problems. Image resolution is 1 m for the optical satellite and 1–3 m for the radar satellites in a program that cost over $2 billion.[199]

Taiwan and South Korea have both built small pre-operational surveillance satellites. Taiwan's Rocsat-2 satellite launched in 2004 provides image resolution of 2 m, while South Korea's Kompsat launched in 2005 provides 1 m imagery. Both were launched on commercially acquired launchers, one American the other Russian. South Korea's next surveillance satellite to be launched in 2009 will be partly funded by the Ministry of Defense and will have 70 cm resolution. The Ministry of Defense is also involved in defining a radar satellite derived from that being developed by Italy (see below) which will have 1 m resolution. Both of these developments reflect the South Korean interest in being able to observe the activities of its neighbor to the north.[200]

GERMANY AND ITALY

During the Cold War, the countries of Western Europe created military forces designed to fight in northern Europe against a Soviet aggressor. Since 1989, however, UN and NATO peace-keeping actions have proliferated, so countries such as Germany and Italy are finding their forces in operation far from Europe. Having

[198] Covault (2007c), and Secretary of State (2006) p. 26.
[199] Morring (2007), and *Space News* (2007).
[200] Taverna (2006b).

Figure 47. The small, low-cost SAR-Lupe imaging radar satellite is Germany's first military surveillance satellite system; the first of a five-satellite constellation was launched on December 19th 2006. Credit: OHB.

depended on US satellite imagery in the past, these countries are now deciding to create a source of imagery independent of American control. France started this trend in the early 1990s (see Chapter 8) with its Helios optical surveillance satellites, and other European countries built on that initiative by forming an image interpretation center in Spain that produced image-based intelligence for any European military force that demanded it. This center was kept particularly busy during the NATO operations in Bosnia and Kosovo dealing with the conflicts that arose as the regions of what had been Yugoslavia splintered into separate countries with associated civil war and ethnic cleansing.

In an example of sensible European burden sharing, Germany and Italy have now both agreed to leave high-resolution optical surveillance satellites to France and to complement them with radar satellites. The sensible nature of this arrangement is spoiled by the fact that Germany and Italy are *each* developing a military radar satellite, not one jointly.

Unlike the billion and a half dollar a throw US radar satellites, Germany will get a fleet of five SAR-Lupe satellites for a total cost of about $400 million (Figure 47). In its spotlight mode of imaging, the 770 kg SAR-Lupe satellite provides 50 cm resolution.[201] The first satellite was launched in December 2006, the second in July 2007, and the remainder should be in orbit by 2008.

[201] De Selding (2004).

The price tag that Italy is paying for its fleet of four Cosmo-SkyMed imaging radar satellites comes to about $1.4 billion.[202] At 1.7 tons each, the satellites themselves are more than twice SAR-Lupe's mass. Both use the same X-band frequency for the imaging radar—this is a higher frequency than most civil radar satellites, making it feasible to get higher resolution. Cosmo-SkyMed will provide imagery in a range of formats and resolutions depending on the application. In the spotlight mode the resolution will be better than 1 m. The satellites are financed jointly by the civilian Italian Space Agency and the Ministry of Defense, so the imagery will be used for both civil and military applications. The first Cosmo-SkyMed was launched in June 2007, with the constellation due to be fully deployed by mid-2009. Argentina is developing two SAR satellites to be launched in the 2010–2012 period that will operate in collaboration with Cosmo-SkyMed. These SAOCOM satellites use the civilian L-band frequency which will make its resolution four to five times worse than that of Cosmo-SkyMed.

Germany is also funding a civilian X-band imaging radar satellite, TerraSAR-X, launched in June 2007, just 8 days after the first Cosmo-SkyMed, which has performance similar to the above two military (or part-military) systems and cost about $250 millon.[203] A follow-on, Tandem-X, costing less than half as much is due to be launched in 2009.

NO MORE COLD WAR SIMPLICITY

From the end of World War II until the collapse of the Soviet Union in 1989–1991, the Cold War dominated global politics. You were on one side of the battle of the social orders—communism versus capitalism—or you were "non-aligned". But your status was determined by where you fitted into the bi-polar picture. Those days are now gone, and for better or worse we live now in a multi-polar world in which yesterday's bad guys are sometimes tomorrow's good guys, and *vice versa*.

Regional tensions take on a greater significance than they did. India–Pakistan, Israel–Arab, Iraq–Iran, North and South Korea, China–Taiwan are some of the flash points that could easily escalate into a regional or even world conflict. The oil dimension has also increased the tension in already sensitive regions such as the Gulf region (the Persian Gulf, that is—not the Gulf of Mexico), the Caucasus/Caspian Sea region, west Africa, and the South China Sea that have too much potential oil for the major powers to ignore their political future.

Surveillance satellites play an important role in keeping these flash points from blowing up into major confrontations. Nuclear weapons have also become part of the equation in several of these hot spots, making the objective intelligence of satellites all the more critical. Where a country on one side of such a zone of tension doesn't have reliable and rapid access to good quality satellite imagery, it is in all our interests that

---

[202] Lardier (2007b).
[203] de Selding (2007b).

the major powers make such imagery available—either providing it from their own satellites or providing a satellite for the country concerned to operate itself.

The Nuclear Non-Proliferation Treaty encourages, arguably even demands, that kind of support to less wealthy countries under Article 1 (don't *induce* a country to acquire nuclear weapons) and Article 6 (general disarmament). The nuclear powers who claim to conform to the Treaty should do more than minimally comply in return for the exemption that the Treaty gives them (the nuclear powers) from the inspection actions of the IAEA.

Because of this implicit requirement for the nuclear countries to pursue dis-armament even more than the non-nuclear countries, the increasing readiness of the last two US Administrations, one Democrat the other Republican, to engage in preventive wars is particularly disruptive. Note that a war is *preventive* if it is not provoked, and I use the word deliberately to distinguish it from the word "pre-emptive" that US and British governments tend to employ. *Pre-emptive* means that an attack by the other party is imminent and can only be forestalled by an immediate attack. The invasion of Iraq was in my view preventive rather than pre-emptive since there was no indication that Saddam Hussein would attack the West, merely that he might develop some particularly nasty weapons. The UK Labour government in the same period has backed the US preventive attack policy. Faced with a possible unprovoked attack from the US and to a lesser extent Britain, non-nuclear states are tempted to gain possession of a nuclear weapon on the grounds that it may *deter* the attacker—nuclear weapons do seem to have worked as a deterrent throughout the Cold War, even if they had no other military utility.

When President Clinton bombed Sudan and Afghanistan in retaliation for terrorist bombings of US Embassies in Africa, he had no time for niceties such as consulting the United Nations before sending in the cruise missiles—the rule of law was too slow for the righteous anger of the victim. To many observers it looked more like an act of war by the US against those countries even to those generous enough not to link Clinton's actions to his personal (Monica Lewinski) abd professional (impeachment) problems. If President George H. Bush (Senior, that is) had the patience to cajole and persuade reluctant nations to back the ousting of Saddam Hussein's armies from Kuwait in 1991—to free a whole country in other words—could not President Clinton's response to an attack on an American Embassy in a far away country wait too?

The second President Bush has tended to favor the Clinton form of international law over that of his father. Preventive attacks have been announced with pride—not the shame that one might expect for paying at best lip service to what little semblance of international law we have. The *war on terror* has become the *war that creates terror*, which is what you would expect when the rule of law is brought into disrepute. The implication of the Non-Proliferation Treaty is that the nuclear powers in return for not being subject to IAEA inspection of their nuclear arsenals should not only be complaint with every nuance of the Treaty, they should be *seen* to be compliant, and the US and Britain in invading Iraq have certainly not met that latter requirement.

The effect of preventive attacks on countries that are out of favor with the US has been to escalate the defense of those countries that fear they may be next on the list—

Iran, for example. There is no doubt that Iran is investing in civil nuclear energy including the facilities needed to provide fuel for the reactors. Whether weapons-grade fuel is being created or is planned is currently unclear—the arguments of the Iranian Ambassador Dr. Ali-Asghar Soltanieh that nuclear weapons are contrary to Iranian religious policy are reasonably persuasive, but it would be foolhardy to rely on his words alone.[204] The intelligence reports that enrichment of uranium is occurring beyond that required for energy production cannot be ignored. However, we recall the intelligence failings about Iraq's weapons of mass destruction (nuclear, chemical, and biological)—these reports underpinned the 2003 invasion of Iraq and turned out to be wrong.[205] The intelligence that underpinned the Iraq "weapons of mass destruction" relied on reports from individuals such as defectors and refugees, and the same is true to some extent in the current Iranian debate. We must remember that defectors don't necessarily want peace; they may want invasion as Hans Blix said (see Chapter 5)—whereas surveillance satellites don't have a hidden agenda.

To continue with the example of Iran, this is a country that lacks its own surveillance satellites, and so rumor of imminent attacks can escalate and lead to unwarranted reactions. The major powers should offer Iran access to good-quality satellite images of US, British, and Israeli forces in the region as a way to defuse tension and to demonstrate good faith. Of course, if they are planning an attack then they are unlikely to go along with this suggestion.

An undesirable but predictable consequence of Iran's toying with nuclear weapons is that other countries in the region are encouraged to do likewise. Countries worried about Iran's zeal in spreading its Shiia form of Islam have begun to seek assistance on nuclear energy from the International Atomic Energy Agency, a development that is seen by many as the first step in a dangerous nuclear arms race in the west Asia region. Countries that have taken this first step include Turkey, Saudi Arabia, Egypt, Syria, and the United Arab Emirates—some of whom have the oil wealth to purchase whatever technology they decide that they need.[206]

The use of a nuclear arsenal to deter a superpower attack re-creates the argument that underpinned the original British and French nuclear bombs—that they alleviated the need to have much larger conventional armed forces to defend against a Soviet attack (i.e., they saved money). Nuclear weapons are also perceived to be a quick and controllable way to deter attack—controllable in the sense that their deployment and use can be more easily controlled by a central decision-maker than is possible with large conventional forces. The 1962 American agreement not to invade Cuba and to withdraw nuclear missiles from Turkey in return for nuclear missiles not being introduced in Cuba shows how this sort of negotiation can work.

---

[204] Cobbold (2007).

[205] At least the US Admnistration had the good sense to replace the Head of the CIA and a number of other top intelligence officials—the UK did the reverse, promoting the man who oversaw key intelligence reports, making him head of the Secret Service (Britain's equivalent to the CIA).

[206] Broad (2007).

The Cuba crisis also shows how horrendously risky such nuclear stand-offs are (see Chapter 3), so it is in every country's interests that this policy is discouraged.

Critics of the preventive attack doctrine argue that a real commitment by the USA, Britain, and other nuclear powers to nuclear weapon non-proliferation would require them to forego the use of preventive attacks in favor of, for example, UN- or EU- or NATO-sanctioned campaigns.

The good news is that most of the countries with a nuclear force also have spy satellites. These satellites allow them to avoid the hair trigger approach to nuclear weapon control, and rely to a great extent instead on unambiguous information about the deployment of its adversary's forces, both conventional and nuclear. Absence of such information would take us back to the hair-raising days of the 1950s when America imagined a huge Soviet missile fleet gearing up to attack, and hawkish American generals proposed preventive attacks on the Soviets as the only way to prevent a communist victory. Thankfully, the first American spy satellites within a few months showed that the Soviets had no more than a handful of cumbersome long-range missiles.

In Chapter 10 we look at the role that spy satellites are likely to play in the global politics of the next few decades.

10

What the future holds

During the Cold War, the main role of spy satellites was strategic—to monitor what the other superpower, and China, were doing as concerns long-range missiles and nuclear weapons. Once the Berlin Wall fell in 1989 and the Cold War slipped into history the priority for these satellites gradually changed—from strategic to tactical.

TACTICAL IMAGERY

With the ending of the Cold War, America found itself drawn into regional conflicts in several parts of the world. The first major example was the Iraq invasion of Kuwait in 1990 which led to the Gulf War of 1991—Operation Desert Shield as the US called it. Satellite imagery was in constant demand by the forces in the Gulf region to assist in their immediate actions—to tell them where enemy forces were located or to clarify the status of infrastructure such as bridges.

But although America had several spy satellites in orbit and collected a great variety of images, the forces on the ground found it difficult to get the information they wanted. Frequently they received loads of images but didn't have the photo-interpreters to work out answers to the questions they faced.

The organizations and infrastructure set up to get satellite images to Washington, DC, where they could be analyzed by specialists supported by banks of super-computers worked well for the Cold War strategic era. Getting information about local conditions to forces scattered across a far away desert was another matter. America spent the 10 years after the first Gulf War addressing this problem, so that by the time of the Iraq invasion in 2003 US forces were better served by the imaging satellite community.

Ironically, the very success of spy satellites has led to expectations often being unfulfilled. The following anecdote illustrates the sort of mismatch between hype and fact that can arise.

A squad commander far from base in Iraq or perhaps Afghanistan calls up his intelligence officer (probably via satellite, incidentally) and asks for an immediate satellite image of the valley ahead of him. The intelligence officer has the wit to query this expensive and probably impossible request (it is highly unlikely that a suitable satellite will be in just the right place at the right time) and asks why the image is needed. The commander explains that he wants to cross the gorge ahead but is worried that the bridge across it has been bombed. The intelligence officer reminds him that even if the satellite image showed the bridge to be down, he might be able to cross if the sides of the gully were not too steep. The commander suggested that this information could be obtained by analysing a satellite image and measuring the slope of the gully sides. The intelligence officer had a better idea—he contacted a special forces unit that he knew was in a position to look down on the valley. The special forces man looked through his binoculars at the gully and reported that cars and trucks were being driven across it by local people, having first bull-dozed a path down the sides. This information was of course exactly what the squad commander wanted to know.

Whereas satellites are excellent at giving the big picture summary of Soviet or American strategic forces, they are less good at giving black and white answers in one particular scene at one particular time. Part of the process of gauging the strategic picture is to examine the changes in a facility or location over time—years, rather than days. The tactical requirements of forces engaged in regional wars are much more immediate and explicit: Is there a tank ahead of me? Is there an enemy soldier in that trench?

The modern way to address those questions is to use unmanned aircraft (scaled-up versions of radio-controlled model airplanes) that take images of the ground below them and radio them back either directly to the forces on the ground or via satellite to a special ground terminal. These drones come in several shapes and sizes. At one extreme they can be operated by the local forces, taking off from a small strip of flat ground, reconnoitering the terrain within a few kilometers of the take-off point, sending images back to a vehicle-mounted terminal, and able to land in the same short strip of flat ground.

At the other extreme is the US Global Hawk vehicle—to give you an idea of its sophistication, the over-run in the budget to develop an enhanced version (just the over-run, mind) is $2 billion—the enhanced version will carry a ton and a half of surveillance equipment instead of just a ton in the current version. This is a serious airplane. It is the size of a small commercial airliner, takes off from an airport runway, and flies for 24 hours non-stop, and over intercontinental distances if required. Global Hawk is not controlled by the man in the front line, nor by the man in the local command center. It is controlled by specially trained staff (a mix of pilots and image analysts) from Beale Air Force Base in California. In the next room is the control center for the U-2 aircraft that still fly reconnaissance missions in Iraq and other regions of interest—the U-2 is still in use because it can carry a more varied suite of sensors than can the Global Hawk at the moment. Images (optical, infra-red, and radar) are beamed to Beale from the craft via satellite or they can be transmitted direct to troops on the ground below. Commands to the craft are sent from Beale to the vehicle via communications satellite.[207]

In between the small, locally operated, unmanned aircraft and the sophisticated Global Hawk are a range of medium-sized vehicles. The Predator is one such, and can not only spy out the lie of the land beyond the next hill, it can do something about it if it finds an enemy there. Unlike the passive Global Hawk which flies at high altitude, Predators are closer to the action and carry missiles that can be fired at a target in its field of view under the control of an operator.[208]

In the example of the squad commander wanting to cross a ravine (see previous page), an unmanned aircraft is more likely to have been the imaging vehicle used— were it needed—than a satellite.

Another form of unmanned aircraft is the cruise missile. These vehicles are not missiles in any technical sense, that's just a name that sounds good and has stuck. They are aircraft, using aerodynamic forces plus jet engines to keep them aloft and propel them through the air. As jet engines have improved, aerodynamic design has become more sophisticated, and lighter weight materials have become available, so the range and load-carrying ability of these vehicles has got better. Some of them can now fly thousands of kilometers, but in practice they are launched a few hundred kilometers from the target. They can be launched from ships, submarines, aircraft (in the air, see Figure 48) or from the ground—usually they are launched from a catapult or squirted out of a tube to give the initial speed to stay airborne.

One feature of current cruise missiles is their ability to navigate unerringly (with the occasional hiccup) to the target—with uncanny precision. The technology to achieve this comprises two main elements—a system for figuring out its position and a digital map. Nowadays a cruise missile figures out where it is with a GPS receiver, like the satnav terminal you buy at the High Street electronics store or perhaps have on your car dashboard. It can also be achieved by the cruise missile monitoring the ground below itself and comparing what it sees with a built-in map in its computer. This is just like the way your in-car satnav system adjusts as you turn a corner or round a bend—it matches your movements to a map and figures out which road you are on by virtue of the twists and turns you are making—a technique called *map matching*. A cruise missile can have at least two types of map to compare with. First is a digital map like the one in your in-car satnav, showing roads, rivers, bridges, buildings, and so on. But the second is a three-dimensional map of the terrain showing the contours of the hills and valleys. These three-dimensional maps are largely created using information from satellites—stereo pictures from CORONA or GAMBIT in the 1960s, for example. Nowadays the 3-D maps can also be the result of mapping using altimeters in satellites. An altimeter measures the height of the satellite above the ground below, so if you can make the altimeter beam narrow enough you can build up a map of the variation in height of the ground below. The cruise missile has its own altimeter with which it observes the terrain below and compares what it sees with the map based on satellite altimetry in its computer

---

[207] Butler (2007b), and Butler (2007c).
[208] Butler (2007d).

Figure 48. A B52 carrying cruise missiles on its under-wing pylons.

memory. Whether the 3-D maps are derived from altimetry, from stereo imaging, or from synthetic aperture radar (SAR) images, satellites have probably been the carrier for the sensor. This technique for cruise missile navigation is called *terrain contour matching*.

For the final phase of its mission some cruise missiles switch to a different form of guidance using infra-red or visible imagery. A camera in the cruise missile takes images of where it expects the target to be and compares the images with pictures in its computer memory. It then adjusts the final moments of its flight to hit the part of the target that has been designated as the aim point. Theoretically, this technique for terminal guidance is so accurate that it can have the cruise missile fly through a window.

Spy satellites remain the only way to get information deep inside another country without the permission of its government and without creating a diplomatic incident. They can detect large-scale troop movements, and thus act as a trip wire to alert neighboring governments to impending attack. The advance warning given by satellite images allows time for diplomatic initiatives to prevent war, or at worst to allow the adversary to prepare its forces to defend against the massing troops. In any case, actions of the governments concerned are taken based on factual evidence, not speculation or rumor.

Satellites also provide information about a country's nuclear developments—not everything you want to know, but quite a bit, and it's objective. They allow you to evaluate electricity generation—a serious constraint in the production of nuclear fuel.

In 1957, for example, the nuclear community consumed 6.7% of all electrical power in the USA; just the two gaseous diffusion plants for uranium enrichment at Oak Ridge, Tennessee, and Hanford, Washington state, consumed more electricity than the entire Tennessee Valley Authority, and the Hoover, Grand Coulee, and Bonneville dams together could have supplied.[209] So, follow the high-tension power cables—that's where you'll find the nuclear facilities.

AN INTERNATIONAL SURVEILLANCE SATELLITE SERVICE

Since correcting the distorted picture of Soviet missile strength in 1961 (the missile gap was in America's favor, not the Soviet Union's as the USAF was claiming), spy satellites have remained a force for objective information on major military dispositions throughout the world. Most of the major powers have access to high-resolution satellite imagery of areas that they designate (they have "shutter control"), and commercial satellites now offer images of the quality that CORONA provided in the 1960s to anyone willing to pay. Is it then time to think of a service to provide strategic imagery for developing countries?

If commercial satellites can supply the required images why is a service needed? One answer was apparent in Afghanistan in 2001 when the US military bought exclusive rights to the best quality images of that region, thus ensuring that no other country (Pakistan or Kazakhstan or Iran, say) could get those images. The service would avoid images with tactical significance but would cover strategic targets such as airports, harbors, power stations, railroad complexes, industrial plant, and the like. In practice, it might be easier to say what is excluded than included. As a starter, the only areas excluded would be UN-sanctioned forces. This definition would have avoided images of the Desert Shield formations massing in the Saudi Arabia desert in 1991 that attacked Saddam Hussein's forces from an unexpected direction.

It need not cost very much. Through the World Meteorological Organization (a UN agency) several countries make weather satellite data available to the world's weather forecasters. The UN doesn't own any satellites or facilities; it just provides a legal umbrella under which individual countries can agree the types of data they will provide in the future, the formats and standards for exchanging data, and so on. Another smaller scale example of the collaborative way to provide a global service is the Disaster Monitoring Constellation created by Surrey Satellite of the UK—satellites are purchased and operated by individual countries and their imagery is made available as required to respond to emergencies such as earthquakes, tidal waves, and famines. Another concept to consider is Google Earth, where images from various satellites are amalgamated and made available to all comers.

For this service to work, an organization would be needed to decide which targets should *not* be imaged, and the IAEA is an example of a possible owner of that task. A regional body such as the European Union could initiate this kind of service as a

---

[209] Rhodes (1995) p. 561.

pilot exercise—to see if there is a demand for it, to explore the technical and political difficulties involved, to see how much it cost, and so on.

NUCLEAR PROLIFERATION

The NPT (Nuclear Non-Proliferation Treaty) seems at times more like the NBBT (Nuclear Back in the Bottle Treaty). With major powers giving mid-sized countries cause to worry about being attacked, it is hardly surprising that some of those mid-sized countries will seek a weapon to deter such attacks. And a nuclear weapon does seem to deter even the superpowers.

The NPT attempts to divide the nations of the world into two groups—those that have nuclear weapons and those that agree never to have them. In return for a country joining the second group, the countries in the first group make several commitments. First of all, the nuclear states agree to help the non-nuclear ones to exploit nuclear power station technology (i.e., for peaceful purposes). They also agree *not* to assist any other country to obtain nuclear weapons. Finally, the nuclear states agree to seek ways to achieve total disarmament—nuclear and conventional. The International Atomic Energy Agency (IAEA) headquartered in Vienna (Figure 49) is tasked with undertaking the various verification agreements in the NPT.

The answer to the proliferation-inducing policy of preventive attacks lies in the hands of the leaders of the major powers. But there are other forces inducing nuclear proliferation that can be tackled at a more mundane level. One particular cause for concern is that terrorist groups might get their hands on enough weapons-grade uranium or plutonium to construct a bomb. Making a bomb out of plutonium is much more difficult than out of enriched uranium (ask the North Koreans about their October 9th 2006 damp plutonium squib), so we can focus on the uranium problem for the moment. A uranium bomb is a much simpler matter. Enriching the uranium

Figure 49. Vienna Headquarters of the UN's International Atomic Energy Agency that polices the Nuclear Non-Proliferation Treaty.

so that it contains a high enough proportion of isotope 235, U^{235}, is a job only governments can undertake, but assembling it into a bomb is a question of having two pieces of the uranium that alone are below critical mass but together exceed that mass. You then fire one into the other like a bullet into a target, and that's a bomb—there are a few other details like the need for a layer of material around the target to prevent neutrons escaping, but all of this information is readily available on the internet. Plutonium is so fissile that a gun is too slow, and by the time the "bullet" has reached the target too many reactions have taken place and the combined mass is no longer critical. But the gun approach works fine for U^{235}.

The likely source of enriched uranium is the 50 tons of it at civilian research reactors around the world—the much greater amount in military centers is thought to be reasonably well guarded against theft by terrorists. The civilian reactors and the associated enriched uranium fuel were mostly supplied by the US and the Soviet Union to about 50 countries during the Cold War—competing to win the support of these countries with an "atoms for peace" program. Security tends to be lax at these research organizations making it possible for pilfering on at least a small scale. Typically the fuel is in a form that can be handled easily, making the security risk all the greater. Besides the fresh fuel for these reactors, another source is the spent fuel that is removed from the reactors. It may be "spent" in terms of fueling a research reactor, but the proportion of U^{235} is still about 80%, more than enough for it to be assembled into a weapon (the IAEA recommends that uranium above 20% enriched be treated as weapons-grade).

Actions are underway funded by the US and a few other countries to retrieve the enriched uranium and to convert the research reactors so that they work with uranium that is not weapons-grade. We must hope that this exercise is completed before the terrorist threat takes advantage of the opportunity.[210]

NUCLEAR ENERGY—GREEN ELECTRICITY, BUT AT WHAT PRICE?

Nuclear weapons, missiles, and surveillance satellites are the triad of hi-tech inventions that dominated the world's dance with death from the start of the Cold War until the 1980s. Just as missiles can be beneficial—for example, when they launch satellites—so too the power of the atom can improve mankind's lot. Indeed, about a sixth of the world's electricity is currently supplied by nuclear power—almost as much as by hydroelectricity. As pressure mounts to reduce carbon emissions, nuclear gains more and more supporters as the fuel of choice.

The safety risks of nuclear reactors have been highlighted by a small number of radioactivity emissions. The worst was the explosion of a reactor at the Chernobyl power plant in the Ukraine in 1986 which spread a cloud of radioactive dust over much of Western Europe—sheep grazing in some parts of Britain 3,000 km away are still unfit for human consumption more than 20 years later. In the US, the 1979 Three Mile Island accident created a crisis of confidence among politicians and the

---

[210] Glaser (2006).

public. Since then, however, safety infringements at nuclear reactors around the world have been minor and infrequent.

If the safety factor is no longer an insurmountable hurdle to building new nuclear power stations, the problem of dealing with radioactive waste may be. Despite a half-century of research, no foolproof method has yet been found for disposing of the proportion of the waste that will remain dangerously radioactive for tens of thousands of years. Burial deep underground is likely to be the method adopted, but there are concerns that even the most geologically stable ground may not be stable over such long periods and thus some of the waste containers may rupture allowing radioactivity to escape. Countries are understandably cautious about leaving such a dangerous legacy to future generations.

Research is however beginning to show how the problem can be tackled, and it is time to press forward with this at top speed. A technique called "advanced closed fuel cycle" avoids creating any significant quantity of the long-life radioactive waste, and consumes existing waste. Still at the pilot plant stage, this technology calls out for an infusion of development funding to scale it up to operational levels.[211]

It seems feasible therefore to build power stations that have a low footprint in carbon, plutonium, and difficult-to-dispose-of radioactive waste.

LOW-COST SATELLITES

At the start of the space age, satellites for surveillance or indeed for pretty much anything were expensive and only governments could consider financing them, but for the past 30 years that has been less and less the case. Of the 600 or so working satellites in orbit today, about 60% are commercial—offering TV, radio, internet access, telecommunications, weather images, and navigation services. The owners of these satellites include some of the hardest nosed business people on the planet who have absolutely no interest in "space glamor" but see satellites as the way to deliver a service with the best return on investment.

Yes, some space programs are still expensive, especially those involving humans in space, and also some military projects. The business of putting humans in space is likely to remain expensive for another decade or two—there are some encouraging signs of a breakthrough in costs but it's too soon to say whether and when human travel into space will become anything other than a high-risk, high-cost adventure.

On the military front, military space projects that were expensive to begin with have incurred eye-watering cost over-runs. The following list of programs gives some idea of the scale of what we are talking about: AEHF (communications) $2.5 billion has risen to $5.8 billion; N-POESS (weather) $6.8 billion is now $8.1 billion, SBIRS (early warning) has risen from $3.9 billion to $10 billion, and FIA (spy satellite) has gone from $6 billion to $10 billion.[212] Back in the 1960s, Senate Minority Leader

---

[211] Hannum (2005).
[212] Cáceres (2006).

Everett Dirksen was said to proclaim *"a billion here, a billion there, and soon you're talking real money,"* and this seems to be the pattern in these military space projects.

Two factors are beginning to cause a re-think in US military space plans—in addition to the spiraling budgets. First, progress in miniaturization and materials has opened up new low-cost options in the past 5 to 10 years. In one sense these technical advances mean that you can cram even more functionality onto a big satellite—and that's the pattern that leads to the big-ticket projects listed above. But another consequence of the better technology is that you can get 80% of the functionality for 20% of the cost—squeezing the last drop of performance out of the latest technology (as is the habit in DoD) inevitably pushes costs sky high; whereas using commercial versions of the technology will provide somewhat lower performance but will be a lot cheaper because of the economies of scale you get from the commercial sector.

In principle, this 80/20 option was always available, but what is different now and will be more so tomorrow is that the performance of commercial electronics, communications equipment, optics, detectors, solar cells, batteries, etc. has reached the point where they deliver useful solutions in orbit. For example, the fact that hundreds of millions of digital cameras are sold each year (including as part of a cell phone) has brought about a revolution in low-cost, light-weight, low power consuming, rugged, small optics and detector technologies. What you need for satellites are optics and detectors that are low-weight, low power consuming, rugged, and small. Ideally, you would like them also to resist radiation and work in vacuum but if the versions that can do that are a thousand or ten thousand times the price, you can find work-rounds. You have to package them more robustly, put them in a sealed box, and so on, but the overall cost–performance trade-off can be quite interesting.

The second factor driving a major re-think in military satellites is the threat of anti-satellite weapons. The Chinese very kindly high-lighted this problem on January 11th 2007 when they destroyed one of their old weather satellites by smashing a missile into it.[213] Technically, this wasn't a particularly difficult feat to accomplish as they had a homing beacon on the target satellite—like a neon sign saying "I'm here". It was certainly an environmentally unfriendly act since it increased the cloud of debris whizzing around in space by at least 10% in one fell swoop.[214] We will come back to why the Chinese may have wanted to create this space spectacular below, but the relevance to the current topic is that it shows the need for owners of military satellites to think about how to survive such attacks in the future.

It is difficult to avoid destruction by a missile that is intent on hitting you in space. The attacking missile doesn't have to do much to knock out a typical satellite, even a military one that has been hardened to survive radiation from a nuclear explosion. If the missile's trajectory is chosen carefully it can hit your satellite at several thousand km/h, which will destroy most objects (this is the tactic the Chinese used in January 2007), but even a slow approach can be fatal if it has the right

---

[213] Covault (2007a).
[214] Morring (2007a).

equipment—zapping the satellite with radiation probably won't work (they are built to survive nasty radiation from the sun, anyway), but spraying gunge on the solar arrays, antenna, or telescope optics will reduce its electrical power, damage its radio link, and fog up its cameras so that it can't do its job.

Your satellite is presumably big and ponderous compared with an anti-satellite weapon, so taking evasive action is unlikely to be a practical solution. Making your satellite invisible has been tried—a stealth satellite, like a stealth bomber—but so far seems not to have worked very well. Ruggedizing your satellite seems like an obvious defence, but can it really be made impregnable to a missile that hits it at thousands of km/h? And being strong and rugged won't protect you from devious tactics like spraying paint on your telescope opening (just like the robbers do to CCTV cameras in the movies).

Seven months before destroying its own satellite, China fired a laser at an American spy satellite that apparently caused no damage. It is not yet clear if this was a test of an anti-satellite weapon or a technique for finding out more about the American satellite. The US has apparently used laser illumination of enemy satellites in the past to gain engineering information about those satellites, and this may have been the Chinese objective.[215]

The only sure tactic against a missile attack like that the Chinese demonstrated is to be able to launch a replacement satellite quickly.[216] And even if you can launch replacement satellites, you won't be happy if the satellites cost $1 billion each. So, the conclusion to be drawn is that instead of big, expensive satellites (compare the 15-ton US Lacrosse with the 300 kg German SAR-Lupe—both radar-imaging satellites) a constellation of small, low-cost satellites is more resilient in the face of anti-satellite weapons. It may take some time for the US and Russia to move away from their traditional no-expense-spared designs, but other countries are heeding the warning signal that China has issued.

CHINA

Let's consider some of the specific issues affecting the main military space powers, and where better to start than with the world's new economic superpower, China.

Since World War II the US has been the hub of the world's economy. Europe and Japan have become major trading partners but have not looked likely to overtake the US in its role as the global agricultural and industrial pace setter. Now in the 21st century will China finally challenge the US?

The up sides are well known. Already the world's fourth largest economy, at current growth rates China will surge past the US in terms of Gross Domestic Product (GDP) before 2020. The trading balance is another eye opener—the US exported $55 billion to China in 2006, but imported a tad more at $288 billion—a trade deficit of $233 billion, up $31 billion from 2005 and the imports include 90% of

---

[215] Covault (2007a).
[216] Muradian (2007).

its rare-earth metals such as terbium used in some fluorescent light bulbs.[217] And China has ploughed its new wealth into better infrastructure, like electrical power and roads. Mind you, not all commentators see China's future as imminent dominance of the world economy; historian David Edgerton remarks that "Most of China's exports, especially in the electronic sector, come from foreign-funded and foreign-owned enterprises, rather than state-owned or locally privately owned ones."[218]

A market economy *and* a communist country, or as Deng Xiaoping said "it doesn't matter if the cat is black or white as long as it catches the mouse." On the military front, China now has the financial muscle to buy or build pretty much anything it wants. As remarked in Chapter 8, China is modernizing its missiles to MIRV status, extending the range of its submarine-launched missiles, and developing cruise missiles. Of the hundred or so missile tests detected by the US DSP satellites each year, about half are said to be Chinese. China's missile production is approaching the rate achieved by the US and the Soviet Union during the 1960s, including four enhanced ICBMs, one of which is road-mobile, a new SLBM, and several new variants of medium- and intermediate-range missiles.[219]

Its spy satellites remain behind those of the West technologically, a legacy of its dependence on the Soviet Union's "keep it simple—and big" designs, and of its lack of access to the latest Western technology (because of military restrictions). Still, being 10 years behind the West means that in five years' time China will have satellites equivalent to the West's commercial satellites five years ago—and these included surveillance satellites with 70 cm resolution.

It may find—as did the US in the 1960s—that the pace of development is limited more by shortages of skilled workers than of funding.

China's main unfinished business is Taiwan 150 km off its southeast shores. When the communists took over China in 1949, only the island of Formosa (now called Taiwan) remained under the control of the beaten Kuomintang or Nationalist Party. Over the ensuing years China has taken over Taiwan's seat on the UN Security Council and established friendly diplomatic and economic relations with America. America though still guarantees to come to Taiwan's defence if China attacks, and China adamantly maintains its claim that Taiwan is an integral part of the Chinese nation. Taiwan blossomed economically long before China itself did, being in the first wave of "Asian tigers" that achieved developed world status in the 1970s. Tentative trade links have been established between the "two Chinas", but resistance to re-integration with the mainland is still strong in Taiwan.

China's recent destruction of one of its own satellites in orbit is seen as somehow connected to its Taiwan ambitions. At the one extreme, it was practice for a campaign to destroy US satellites when China decides to invade Taiwan. At the other extreme, it was a wake-up call to the US to get serious about banning weapons from space. Or perhaps it was a bit of both—achieve international agreement to ban space

[217] Coy (2007), and Cohen (2007).
[218] Edgerton (2006) p. 137.
[219] Covault (2007d).

weapons, thus putting a stop to US developments of that type, but be ready to blast US satellites out of the sky if and when necessary in the future.[220]

An armed initiative against Taiwan seems unlikely in the near term—not until China has exhausted economic-based attempts to regain the island. The US trade imbalance with China has now reached a scale that gives China considerable influence over America's economy. If China were to start to sell off its US dollar holdings, the dollar would drop in value. Of course, such a tactic would hurt China as well as the US, so the first tactic is likely to be a more subtle threat, perhaps no more than a request to encourage Taiwan to think seriously about rejoining the mainland. US pressure on Taiwan to negotiate a political linkage with China would be difficult for Taiwan to resist, since it depends heavily on the US for a defensive umbrella against Chinese military blackmail.

If a China–Taiwan interaction were to get close to a shooting war, China will no doubt feel that its nuclear force will deter radical US actions such as invasion or strategic bombing. China will also presumably see its nuclear force as a useful deterrent in any future disputes with Russia over oil—already a source of tension in the Caspian Sea region.

More generally, China has largely avoided military operations against foreign countries—I am being generous to China here in designating Tibet as part of China rather than a foreign country. However, if its wealth continues to increase, it may become tempted to project its influence farther afield using its military assets, in the way that the UK did in the 19th century and the US in the 20th.

This force projection temptation may also afflict other giant, emerging, economic power-houses—Russia and India, for example. The need for other countries to have surveillance satellites to monitor potentially provocative Chinese or Russian or Indian or other mini-superpower military assets far from the homeland will increase.

THE USA—NEXT STEP: MISSILE DEFENSE?

The spy satellites planned by the US for the next two decades are intended to provide tactical support to US forces around the world—and continue to monitor strategic targets such as ICBMs, submarine construction, nuclear fuel enrichment, and the like. The tactical users want high-resolution radar imaging all the time and with special functions like stereo imagery and detection of moving targets. They want the adversary to have no opportunity to move during a gap between satellites—which can happen now even though the US has eight optical and radar satellites in orbit.

The Future Imagery Architecture (FIA) and the Space-based Radar (SR) are the answer to this requirement, except that their price tag is probably unaffordable—estimates in the trade press run from $10 billion up to five times that.[221] The competition from unmanned aircraft mentioned above is another reason to hesitate before committing to these ambitious programs. The first FIA satellites might appear

---

[220] Perrett (2007).
[221] Cáceres (2006), and Ferster (2007).

in 2009 or 2010, and those would probably only be prototype systems. Meanwhile, four of the eight Advanced KH-11 (optical) and Lacrosse (radar) imaging satellites were launched before 1997 and are thus well beyond their 7–8 year design lives. Two more were launched before 2002, so by 2010 the eight could well be down to two— and there are no more under construction.[222] The US may be forced to build and launch some small, rapidly constructed optical and surveillance satellites to fill the gap between the demise of its current fleet and the availability of the next-generation FIA satellites. Perhaps that stop-gap would then become a 10-year interim solution— just like the CORONA capsule return solution in 1960 stepped in "temporarily" for more than a decade while work continued to develop a radio-link version.

The second big spy satellite investment underway in the US is for defense against ballistic missiles. The *star wars* Strategic Defense Initiative (SDI) in the 1980s may have persuaded the Soviets that they couldn't afford the cost of countering SDI in what appeared to be a new round of escalation in the missile arms race, thus leading to the end of the Cold War. But SDI was more spin than substance, and seems unlikely to have ever worked as advertised: an impervious shield against Soviet missile attacks. US ambitions are now more modest, namely to defend against one or two missiles from a "rogue state" such as Iran or North Korea. It avoids SDI's orbiting laser battleships and dozens of surveillance satellites, and instead relies on a few satellites to detect missile launches (the so-called early-warning satellites described in Chapter 5), powerful ground-based radars, and fast ground-based interceptor hit-to-kill missiles located under the expected flight path of the attacking missiles. Missiles are already deployed in Alaska to intercept a North Korean attack and sites are now being sought in eastern Europe to deal with an Iranian attack— Poland for the interceptor missiles and the Czech Republic for the radar are tipped as likely locations. More ambitious concepts to shoot down a missile just after its launch (when full of fuel and traveling relatively slowly) are still under development, while at the other end of the flight path two anti-missile systems are deployed, one on US Navy ships, the other on land.[223] While these various systems have (after many attempts) been shown to work against test targets, their effectiveness in a real engagement remains to be demonstrated. In the 1960s it was considered that defense against missiles was five times the cost of the missiles being defending against,[224] so the advantage always lay with the attacker. There is no agreement on whether that equation has changed to favor the defender or the attacker—arguments can be presented to support either contention—so the jury is still out on whether missile defense is affordably feasible.

The space element of this US defensive shield is the evolutionary successor of the DSP described in Chapter 5. Sensors on the satellite detect the flash of a missile launch and cue the ground-based radars that direct the interceptors. Since the first was launched in 1970, 20 of the 2.5-ton DSP satellites have been placed into a highly elliptical orbit that gives them long viewing times over the areas of interest. The next

---

[222] Covault (2006a).
[223] Robinson (2007).
[224] DeGroot (2005) p. 299.

generation of early-warning satellites is intended to comprise a constellation of about 20 satellites in a low orbit plus 4 satellites in high orbits (2 in geostationary orbit and 2 more in highly elliptical orbits). The high satellites spot the bright rocket flame as the missile accelerates (the high-elliptical orbits give coverage of launches from northern polar regions), then the low-orbit satellites are intended to track the missiles after their rocket motor has burned out and they enter the post-boost and coast phases of their journey. Although the satellites in low orbit are relatively small (half a ton or so) and thus not too expensive, the sheer numbers required make the concept an expensive one and it is not expected to be operational until 2010 or beyond—the first demonstrator satellite is scheduled for launch from Cape Canaveral in November 2007. The first demonstrator of the highly elliptical part of the constellation was launched on June 27th 2006 and was considered to have been a success, and the first geostationary demonstrator is due for launch in late 2008, after which a decision on how to proceed will be taken making deployment of the operational versions likely from 2010 onwards. The high-orbit part of this new missile detection system has been dogged by schedule delays and budget rises, currently forecast to be $11 billion. Looking even further ahead, research is already underway on the successor to the high part of the constellation, currently dubbed Alternative Infrared Satellite System (AIRSS) involving advanced detectors to provide more accurate tracking than the detector systems currently used.[225]

Other countries in the missile defense business include Russia whose nuclear-tipped ABM system has been deployed around Moscow for nearly 40 years, Israel whose plans in this domain were outlined in Chapter 9, and Europe which is enhancing air defense systems to cope with tactical ballistic missiles like the Scud. A recent entrant is India which demonstrated the ability to intercept an inbound ballistic missile in November 2006, a hit-to-kill feat hitherto only demonstrated by the US and Israel. Further tests are planned of two versions of the system—one to intercept above 50 km, the other below 30 km. Entry into operational service is scheduled for 2012.[226]

Of course, as mentioned in Chapter 9, the US might reduce the threat of missile attack more cost-effectively than by anti-missile systems by refraining from preventive attacks on nations with which it is in dispute.

INDIA—REGIONAL SUPERPOWER FOR THE INDIAN OCEAN?

India's impressive anti-missile achievement and plans are part of an ambitious program of investment in space and strategic programs. In economic terms India is today where China was about 10 years ago. Business is booming but the country's infrastructure is creaking at the seams—some would say has burst right through. China invests about 9% of GDP in public works compared with 4% for India, which

---

[225] Butler (2007a), Butler (2006), Cáceres (2006), Richelson (1999) pp. 226–226, and Lardier (2007a).
[226] Raghuvanshi (2007).

suggests that China will continue to pull ahead unless India boosts infrastructure investment soon. China's export performance also outshines India's—China's exports make up 7% of global trade, India's less than 1%. India lags in other important indicators including energy production (a quarter of China's), freeways (one-seventh), and internet penetration (a third), but hardly at all in population (1.1 billion compared with China's 1.3 billion). In recognition of the widening of the economic gap with its great Asian rival, the Prime Minister has promised up to $500 billion in new highways, power generation, ports, and airports over the next five years. Given the high level of public debt (82% of GDP, one of the highest in the world) these high-priority investments are likely to threaten government investment in other areas.[227]

One of the areas of investment at risk is space programs that are primarily prestige-oriented rather than addressing down-to-earth applications. The defense budget is probably immune to this effect, so missile defense, submarine-launched ballistic missiles, aircraft carriers, and other force projection systems will probably continue to be funded. But plans to put humans in space and to land probes on the moon must be candidates for review. The human spaceflight program is a response to China's launch of Taikonaut Yang Liwei in 2003, making some Indians feel relegated to a lower rank in the space hierarchy, but whether the Indian half of the new Asian space race will get the necessary funding remains to be seen.

BRITAIN AND FRANCE

The two purely European nuclear powers (Russia being Eurasian) are Britain and France. Both countries publicly urge other countries to respect the Nuclear Non-Proliferation Treaty (NPT). Some would say, however, that their commitment to upgrade their own nuclear forces tends to undermine these entreaties to other states not to seek nuclear weapons. Both countries argue that their stated intention of reducing the number of submarines and missiles in their armory fulfills their commitment under the NPT to work towards disarmament.

As discussed in earlier chapters, France is active in providing assistance to other countries in civil nuclear energy as required by the NPT, while Britain has lost much of the capability to do so.

France and Britain justify retention of a nuclear capability by arguing that the world is an uncertain place, and that one cannot rule out the need for a nuclear deterrent 40 or 50 years from now. This argument underpins the decision on both sides of the English Channel to keep their nuclear forces in operational condition for the foreseeable future. Neither country explains why other countries shouldn't use the same argument to justify possessing the Bomb.

As discussed in earlier chapters, France has a mature spy satellite program, Helios, with two satellites in orbit, and has agreements with Germany and Italy to have access to data from their radar-imaging satellites. A new generation of the

[227] Hamm (2007).

French commercial imaging SPOT satellites, called Pleiades, will comprise two satellites that are smaller and cheaper than the current system, and will have 50–70 cm resolution that gives them a military significance. Originally planned for launch in 2008, the first launch is not now planned until late 2009. Plans for a follow-on to the SPOT medium-resolution satellite are under discussion.[228] Future versions of Helios are likely to follow the same trend to smaller satellites, and more of them.

Britain has followed a different path in surveillance satellites, relying on US resources for strategic and tactical imagery. Britain has, however, sponsored the emergence of a native industrial capability in commercial surveillance satellites, resulting in sales of satellites to several countries including Nigeria, Algeria, and China. The UK Ministry of Defence (MoD) has also funded a demonstrator of a low-cost military satellite, Topsat, built by a consortium of three British organizations[229] (Figure 50). The MoD is reviewing the requirements for what it calls "deep and persistent surveillance" together with what it has learned from using Topsat experimentally in Iraq and Afghanistan before deciding whether to develop an operational space-based surveillance system.[230] The trade-offs with other options (unmanned aircraft, for example) are being undertaken through the DABINETT program.[231] More generally, small satellites for reconnaissance have been designated by the UK MoD as a priority area in which to maintain a national capability.[232]

Britain does have a strong heritage in space-based imaging radar. Having been home to the invention of radar in 1935, the UK led the development of Europe's first satellite-borne imaging radars—on the civilian ERS-1, ERS-2, and Envisat satellites. The UK military have presumably had access to US radar imaging, and thereby built up skills in extracting information of military significance from such images. Objects look subtly different in a radar image than in a visible or infrared image; shadows take on a different meaning in radar imaging, for example, because the illumination comes not from the sun but from the radar. As noted in Chapters 8 and 9, several countries are launching satellites with imaging radars that have sufficiently powerful resolution to be of military use. Thus, there will be a demand by countries that receive or buy those images for assistance in reliably extracting information from them—a demand that the UK is probably best qualified in Europe to satisfy.

Recent policy statements by both Britain and France have provided no rationale why Western Europe needs two nuclear powers each with separate missile and submarine delivery systems.[233] Appeals from Britain and France to non-nuclear countries to refrain from joining the nuclear club would presumably carry more weight if one or both were to give up its deterrent. Since both countries have invested tens of billions of dollars in creating national capabilities in the relevant technologies there would obviously be some hard bargaining before agreeing a merger of their

---

[228] Taverna (2006a).
[229] QinetiQ, Surrey Satellite Technology and Rutherford Appleton Laboratory.
[230] Select Committee (2006) pp. 304–305.
[231] Prime contractor: LogicaCMG.
[232] Secretary of State for Defense (2005) p. 110.
[233] DICoD (2007), and Secretary of State (2006).

Figure 50. Image of the Dartford road bridge over the Thames Estuary near London taken by the small, low-cost ($25 million) Topsat satellite, launched October 27th 2005. Credit: TopSat Imagery © QinetiQ 2006.

strategic capabilities. Nevertheless, if the political will existed to make it happen, such a merger would send a clear signal to the non-nuclear world that Europe was serious about non-proliferation, and would still leave Europe holding a nuclear card.

In any sensitive negotiation it is important to have some bargaining cards with which to influence the outcome. The current duplication of nuclear deterrent in France and Britain gives Europe one such major bargaining chip in discussions with, for example, Iran. Historically, Britain is perceived as the foreign power with the worst record of meddling in Iranian affairs—much more so than America.[234] Thus, an offer in which Britain gives up significant international prestige would be looked

[234] Tait (2007).

on favorably in Tehran. Therefore, one diplomatic tactic that might succeed in persuading the Ayatollahs to forswear nuclear weapons is for Britain and France to agree to reduce the number of European nuclear powers to one. This bargaining card can only be used once if it succeeds, so it is important to get the maximum in return for playing it. Another bargaining card of similar weight would be an offer for Britain and France to merge their two Permanent Seats on the UN Security Council into a single European one. Again, this can only be done once, so the benefit has to be commensurate with the high political price.

The same consideration applies to the details of any Franco-British merged nuclear deterrent. If, for example, Britain were to state now that it was reluctant to replace its Trident missiles when they become obsolete, it would lose all bargaining power in negotiations with France on such a merger. France is in the middle of a major investment to replace its equivalent for Trident—the M51 replacing the M45. Hence until the UK commits funding to a Trident replacement the French are unlikely to even consider a shared submarine-based missile fleet. This argument has not been mooted during the recent UK Parliamentary debates on the Trident replacement, but perhaps it should have been.

Another complaint from non-nuclear states about Britain and France is that the size of their nuclear arsenals is bigger than can be justified by any conceivable threats. The nuclear deterrent works by threatening the destruction of a major installation in the adversary's country. The Cold War scenario where Britain or France threatened to destroy several tens of Soviet installations is no longer realistic because that will create a world-threatening amount of radioactive fallout. And by the way, Russia is now the main supplier of natural gas to Western Europe, so a threat to ravage it with nuclear bombs is like a threat to cut off your own arm and/or leg. A credible deterrent needs therefore only a handful of bombs ready to retaliate at any time. The 160 (UK) or 350 (France) bomb arsenals and the roughly 50 warheads in each submarine are massively over-kill for this purpose.

Thus, another step for Britain and France to take that would demonstrate a commitment to nuclear disarmament would be to develop smaller missile-carrying submarines consistent with the idea of minimal deterrence. Such submarines make sense too in terms of allowing two or three to be on station at any given time, thus avoiding the syndrome of "all eggs in one basket" if, for example, one has an accident or is discovered by the anti-submarine detection equipment of an adversary.

The idea that a nuclear bomb leads to reduced conventional defence spending has been consigned to the dustbin of history for more than 50 years. Another lesson from 20th century history is that a massive nuclear arsenal is no more of an assured defence against a clever adversary than was the Maginot Line in France in 1939.

TRUST THE INTELLIGENCE

Strategic intelligence is headline news these days. North Korea has a nuclear bomb— or does it? Iran is enriching uranium—or is she? Iraq had weapons of mass destruction ready to fire with 45 minutes' notice—actually, no she didn't. These pieces of information, whether right or wrong, lead to wars with horrendous consequences.

Some people say that we must have *trust* in order to make the world a safer place. I find that notion a bit naive having lived through half of the 20th century—the bloodiest century in history, the century of state-sponsored mass murder. Far better to arrange matters so that we can have *confidence* that the actions signed-up to by the other party to an agreement will in fact occur. During the Cold War America and the Soviet Union had to invent a new form of diplomacy to cope with a situation where war with the other party would destroy the world. By and large we can see in retrospect that they did quite a good job. Let's therefore remember the lessons they learned as they groped their way towards a viable international order.

High on the list was the need for reliable *verification* of the military readiness of the other party. Conventional sources of intelligence information led to, first, the non-existent nuclear arsenal of the US in the years immediately after World War II which caused the Soviet Union to develop the atomic and H-bombs. Next was the non-existent bomber gap that triggered a rapid build-up of US strategic bombers. Then came the non-existent missile gap influencing the outcome of a US Presidential election and almost triggering a massive US missile deployment. Almost—but not quite, because surveillance satellites arrived just in time to show that the missile gap did exist but it was the other way round: America had the lead in missiles not the Soviets.

Spy satellites rapidly became the reference source of information on strategic missiles and weapons—providing unambiguous, authoritative, and surprisingly comprehensive details of developments around the world. They allowed the super-powers to engage in debate about how to limit their ridiculously overstocked nuclear arsenals, with the *confidence* that any violation of an agreement would be detected. The need for confidence led to the resulting Treaties being formulated so that satellites *could* verify their terms—they put limits on silos not on missiles, on bombers not on bombs, and on tested systems not on theoretical ones.

The 21st century has all the makings of a peaceful and prosperous period for humanity—or of another round of mass killings and misery, with weapons even more horrific than those of the last century. We must encourage our leaders to avoid unnecessary increases in tension and to build confidence between nations and regions. Decisions to wage war based on ambiguous intelligence do *not* build confidence.

Surveillance satellites reduce the occurrence of military false alarms, and they provide the factor of time that eliminates hair-trigger control of weapons of mass destruction. This resource should be made available to all nations that feel threatened militarily so as to give them reliable information on which to make military decisions, not base them on rumor and speculation.

References

Air & Cosmos (2006) ÉTATS-UNIS Big Brother is watching you, February 3rd, **2016**, 8.

Ambrose S. E. (1984) *Eisenhower: The President 1952–1969* (London: George Allen & Unwin).

Aspin L. (1979) The Verification of the SALT II Agreement, *Scientific American*, February, **240**(2), 30–37.

Barrie D. (2006) Going Ballistic—Longevity will determine at which point UK opts for a successor warhead for next-generation Trident missile, *Aviation Week & Space Technology*, December 11th, 34.

Barrie D. and Butler A. (2006) Chevaline Redux—Desire for national strategic capability has UK seeking Trident follow-on input, *Aviation Week & Space Technology*, November 27th, 29.

Blair B., Feiveson H. A., and von Hippel F. N. (1997) Taking Nuclear Weapons off Hair-Trigger Alert, *Scientific American*, November, **275**(5), 42–49.

Broad W. J. and Sanger D. (2007) With Eye on Iran, Rivals also Want Nuclear Power, *New York Times*, April 15th, Late edition—Final, **1**, 1.

Burrows W. E. (1986) *Deep Black: Space Espionage and National Security* (New York: Random House).

Butler A. (2006) Fresh Eyes—First images from new Sbirs sensor bring hope for the program after major setbacks, *Aviation Week & Space Technology*, November 20th, 22–23.

Butler A. (2007a) Mil$space—Space Control sees small piece of the Air Force's white budget, *Aviation Week & Space Technology*, February 12th, 26–27.

Butler A. (2007b) Constantly Watching—the Global Hawk is coming of age in Oraq, Afghanistan, *Aviation Week & Space Technology*, March 12th, 56–59.

Butler A. (2007c) Going Global—remote piloting allows USAF to consider centralized high-altitude recon ops from California, *Aviation Week & Space Technology*, March 12th, 60–61.

Butler A. (2007d) Caught on Tape—insurgents shift tactics in Iraq to try to evade Predator sensors, *Aviation Week & Space Technology*, March 12th, 62.

Cáceres M. (2001) Orbiting satellites: bean-counter's heaven, *Aerospace America*, August, 22–26.

Cáceres M. (2006) Cost overruns plague military space programs, *Aerospace America*, January, 18–23.

Carter B. (1974) Nuclear Strategy and Nuclear Weapons, *Scientific American*, May, **230**(5), 20–31.

Chang J. and Halliday J. (2005) *Mao: The Unknown Story* (London: Jonathan Cape).

Clark P. (2001) Russia has no reconnaissance satellites in orbit, *Jane's Defence Weekly*, May 8th.

Cobbold R. (Chairman), Soltanieh A.-A., Hoare J., and Jahanpour F. (2007) How Might a UK Decision to Replace Trident Affect Key Current Proliferation Concerns? *Regional and Global Perspectives RUSI-ORG Joint Conference, March 16th, London*.

Cohen D. (2007) Earth audit, *New Scientist*, May 26th, 34–41.

Collins G. P. (2007) Kim's Big Fizzle—the physics behind a nuclear dud, *Scientific American*, **296**(1), 8–9.

Covault C. (2006a) Fade to Black—US secret satellites make 16 runs a day over Iranian nuclear sites, but such comprehensive intel may not last the decade, as space recon dwindles, *Aviation Week & Space Technology*, May 15th, 24–26.

Covault C. (2006b) Israeli Overhead—Striking images from new spacecraft also send messages to Syria and Iran, *Aviation Week & Space Technology*, May 15th, 26–28.

Covault C. (2007a) Space Control—Chinese anti-satellite weapon test will intensify funding and global policy debate on the military uses of space, *Aviation Week & Space Technology*, January 22nd, 24–25.

Covault C. (2007b) Iran's Sputnik—Tehran looks poised to try satellite launch with long-range missile implications, *Aviation Week & Space Technology*, January 29th, 24–26.

Covault C. (2007c) Volatile Mix—Iran–North Korean missile collaboration grows as covert Chinese Asat possibility lingers, *Aviation Week & Space Technology*, March 5th, 24–26.

Covault C. (2007d) Eyes on China and Iran, *Aviation Week & Space Technology*, April 9th, 48–54.

Coy P. (2007) The Trade Gap—There's Good News Too, *Business Week*, February 26th, 13.

Day D. A. (2003) US Government Declassifies Reconnaissance Satellites Information, *Spaceflight*, **45**, 116–117.

Day D. A., Logsdon J. M., and Ladell B. (eds.) (1998) *Eye in the Sky* (Washington, DC: Smithsonian Institution Press).

De Groot G. (2005) *The Bomb—A History of Hell on Earth* (London: Pimlico).

De Groot G. (2007) *Dark Side of the Moon: the Magnificent Madness of the American Lunar Quest* (London: Jonathan Cape).

de Selding P. (2004) German Military Prepares for 2005 SAR-Lupe Deployment, *Space News*, May 24th, 6.

de Selding P. (2007a) French Minister Urges 50% Increase in Military Space Spending, *Space News*, February 19th.

de Selding P. (2007b) Germany's TerraSAR-X launched, *Space News*, June 18th, 4.

Deutch J. M. and Moniz E. J. (2006) The Nuclear Option, *Scientific American*, September, **295**(3), 52–59.

DICoD (Délégation à l'information et à la communication de la Défense) (February 2007) *Let us Make more Space for our Defence—Strategic Guidelines for a Space Defence Policy in France and Europe*, ISBN 2-11-096440-5.

Drell S. D. and von Hippel F. (1976) Limited Nuclear War, *Scientific American*, November, **234**(5), 27–37.

Dupont J. (2006) Dissuasion: le M51 entame ses essais—Un premier tir sans faute qui valide les options prises en 1996 d'un programme d'essais restreint, *Air & Cosmos*, **2053**, November 17th, 24–26.

Dupont J. (2007) La France rénove son arsenal nucléaire, *Air & Cosmos*, **2082**, June 15th, 96–98.

Edgerton D. (2006) *The Shock of the Old: Technology and Global History since 1900* (London: Profile Books).

Epstein W. (1975) The Proliferation of Nuclear Weapons, *Scientific American*, April, **232**(4), 18–33.

The FAS site is at *http://www.fas.org/irp/imint/resolve5.htm*

Ferster W. (2007) CBO Report Highlights Limitations of Space Radar Options, *Space News*, January 29th, 13.

Gaddis J. L. (2005) *The Cold War* (London: Allen Lane).

Garwin R. L. (1972) Antisubmarine Warfare and National Security, *Scientific American*, July, **227**(1), 14–25.

Glaser A. and von Hippel F. N. (2006) Thwarting Nuclear Terrorism, *Scientific American*, February, **294**(2), 38–45.

Gorin P. A. (1997) Zenit—the first Soviet photo-reconnaissance satellite, *Journal of the British Interplanetary Society*, **50**, 441–444.

Gorin P. A. (1998) Black 'Amber': Russian Yantar-Class Optical Reconnaissance Satellites, *Journal of the British Interplanetary Society*, **51**, 309–320.

Greenwood T. (1973) Reconnaissance and Arms Control, *Scientific American*, February, **228**(2), 14–25.

Hamm S. (2007) The Trouble with India—Crumbling roads, jammed airports, and power blackouts could hobble growth, *Business Week*, March 19th, 49–58.

Hannum W. H., Marsh G. E., and Stanford G. S. (2005) Smarter Use of Nuclear Waste, *Scientific American*, December, **293**(6), 64–71.

Harford J. (1997) *KOROLEV: How One Man Masterminded the Soviet Drive to Beat America to the Moon* (New York: John Wiley & Sons).

Healey D. (1989) *The Time of My Life* (London: Michael Joseph).

Isaacs J. and Downing T. (1998) *Cold War* (London: Transworld Publishers).

Johnson P. (1983) *Modern Times: The World from the Twenties to the Eighties* (New York: Harper & Row).

Jung P. (2007) Tally of satellites placed in orbit, *Journal of the British Interplanetary Society*, **49**, 35.

Kennedy P. (1989) *The Rise and Fall of the Great Powers* (London: Fontana Press).

Kislyakov A. (2006) 108 Minutes that Changed the World, *Spaceflight*, **48**, 229–231.

Kissinger H. (1979) *The White House Years* (New York: Weidenfeld & Nicolson and Michael Joseph).

Kissinger H. (1982) *Years of Upheaval* (Boston: Little, Brown & Company).

Kissinger H. (1999) *Years of Renewal* (London: Weidenfeld & Nicolson).

Langewiesche W. (2005) The Wrath of Khan—How A. Q. Khan made Pakistan a nuclear power and showed that the spread of atomic weapons can't be stopped, *The Atlantic*, **296**(4), 62–85.

Lardier C. (2007a) Le budget spatial militaire américain 2007, *Air & Cosmos*, February 17th, **2018**, 36.

Lardier C. (2007b) Succès pour Cosmo/Skymed-1, *Air & Cosmos*, June 15th, **2082**, 152.

Lardier C. (2007c) Donner plus d'espace à la Défense et à l'Europe, *Air & Cosmos*, June 15th, **2082**, 100–105.

Lindgren D. T. (2000) *Trust but Verify* (Annapolis, MD: Naval Institute Press).

McDonald R. A. (ed.) (2002) *Beyond Expectations—Building an American National Reconnaissance Capability: Recollections of the Pioneers and Founders of National Reconnaissance* (Washington, DC: American Society for Photogrammetry & Remote Sensing).

Mikhailov V. N. (ed.) (1999) *Catalog of Worldwide Nuclear Testing* begell-atom (*http://www.iss.niiit.ru/ksenia/catal_nt/intr.htm*).

Morring F. Jr. (2007a) Worst Ever—Chinese anti-satellite test boosted space debris population by 10% in an instant, *Aviation Week & Space Technology*, February 12th, 20–21.

Morring F. Jr. (ed.) (2007b) Watching North Korea, *Aviation Week & Space Technology*, March 5th, 15.

Muradian V. (2007) Test Heightens U.S. Concerns About China's Space Strategy, *Space News*, February 5th, 4–5.

Myrdal A. (1974) The International Control of Disarmament, *Scientific American*, October, **231**(4), 21–33.

Norris P. (1994) *Flight Mechanics Issues Encountered on the First Apollo Lunar Landing Mission*, AAS 94-135 (Springfield, VA: American Astronautical Society).

Opall-Rome B. (2007) Israel MoD Panel to Reaffirm Choice of Rafael Missile Shield—Concept for Defeating Short-Range Rockets Wins Out Despite Fierce Lobbying by Competitors, *Space News*, January 29th, 12.

Perrett B. (2007) Asat strategy—Destroying satellites may be a capability China doesn't want, *Aviation Week & Space Technology*, January 29th, 26–27.

Raghuvanshi V. (2007) India Plans Second Anti-Ballistic-Missile Test in June, *Space News*, January 29th.

Richelson J. T. (1991) The Future of Space Reconnaissance, *Scientific American*, January, **264**(1), 18–24.

Richelson J. T. (1999) *America's Space Sentinels: DSP Satellites and National Security* (Lawrence, KS: University Press of Kansas).

Richelson J. T. (2002) *The Wizards of Langley: Inside the CIA's Directory of Science and Technology* (Boulder, CO: Westview Books).

Robinson T. (2007) Fielding Missile Shields, *Aerospace International*, March, 30–33.

Scoville H. Jr. (1971) The Limitation of Offensive Weapons, *Scientific American*, January, **224**(1), 15–25.

Scoville H. Jr. (1972) Missile Submarines and National Security, *Scientific American*, June, **226**(6), 15–27.

Secretary of State for Defence (2005) *Defence Industrial Strategy* (London: HMSO, Cm6697).

Secretary of State for Defence and Secretary of State for Foreign and Commonwealth Office (2006) *The Future of the United Kingdom's Nuclear Deterrent* (London: HMSO, Cm6994).

Select Committee on Science and Technology (2006) *Written Evidence Submitted to Inquiry on Space Policy* (London: House of Commons).

Siddiqi A. A. (2003) *Sputnik and the Soviet Space Challenge* (Gainesville, FL: University Press of Florida).

Simpson S. (2007) Seismic Sentries—why underground nuclear tests are so hard to hide, *Scientific American*, **296**(1), 9–10.

Smart S., Heavens A., and Frenk C. (2007) PAN-STARRS to survey the Universe, *Frontiers*, **26**, 12 [ISSN 1460-5600].

Soares C. (2005) Cold War Clues—atomic tests allow carbon dating of baby boomers, *Scientific American*, December, **293**(6), 11–12.

Space News (2007) Japan Lofts Two Imaging Satellites for Constellation, *Space News*, March 5th, 4.

Spufford F. (2003) *Backroom Boys—the Secret Return of the British Boffin* (London: Faber & Faber).

Tait R. (2007) Why Iran took its fight to the open seas, *The Observer* (London), March 25th, 32.

Taverna M. A. (2006a) Mix and Match—High-res sensors to reinforce Spot Image product line and revenues, *Aviation Week & Space Technology*, November 20th, 24–25.

Taverna M. A. (2006b) Going Submetric—Taiwan, South Korea expand remote sensing networks and satellite design expertise, *Aviation Week & Space Technology*, November 20th, 25.

Taverna M. A. (2007) Under the Wire, *Aviation Week & Space Technology*, January 22nd, 33.

Temple L. P. III (2005) *Shades of Gray: National Security and the Evolution of Space Reconnaissance* (Reston, VA: American Institute of Aeronautics and Astronautics).

Thomas H. (1971) *Cuba: or the Pursuit of Freedom* (London: Eyre & Spottiswoode).

Thompson T. (ed) (1997) *TRW Space Log 1996* (Redondo Beach, CA: TRW Inc.).

Tsipis K. (1975) The Accuracy of Strategic Missiles, *Scientific American*, July, **233**(1), 14–23.

Turnill R. (2003) *The Moonlandings—an eyewitness account* (Cambridge, UK: Cambridge University Press).

Union of Concerned Scientists (2007) *UCS Satellite Database* (updated regularly) (*http://www.ucsusa.org/global_security/space_weapons/satellite_database.html*)

Uranium Information Centre (March 2007) *Briefing Paper 28—Nuclear Power in France* (Melbourne, Australia).

Ward P. D. and Brownlee D. (2000) *Rare Earth: Why Complex Life is Uncommon in the Universe* (New York: Copernicus).

Webb S. (2002) *Where is Everybody?* (New York: Copernicus Books).

Werth A. (1965) *Russia at War* (London: Pan Books).

Wilson A. (ed) (1995) *Jane's Space Directory Eleventh Edition 1995–96* (Coulsdon, UK: Jane's Information Group).

Wolfe T. (1980) *The Right Stuff* (New York. Bantam Books).

York H. F. (1973) Multiple-Warhead Missiles, *Scientific American*, November, **229**(5), 18–27.

York H. F. (1975) The Debate over the Hydrogen Bomb, *Scientific American*, October, **233**(4), 106–113.

York H. F. (1983) Bilateral Negotiations & the Arms Race, *Scientific American*, October, **249**(4), 105–113

Index

(**boldface** indicates main entry page range)